A VIEW OF MANAGEMENT IN FIRE INVESTIGATION UNITS

VOLUME I

**Issues & Trends
for the 90's**

March 1990

This document produced by Tri-Data Corp. under Contract Number EMW-88-R-2869. The opinions expressed in this document do not necessarily reflect the positions of the United States Fire Administration or the Tri-Data Corporation.

Table of Contents

Acknowledgements

The U.S. Fire Administration and TriData wish to thank the many individuals who contributed to the research for this project. Mr. Tom Minnich, Project Officer, spearheaded this important research, and ensured that the project produced information that could be shared with any individual or organization concerned about fire investigation management and arson control. Hollis Stambaugh was the project manager for TriData.

Mr. Daniel Carpenter and Mr. Randy Kirby served as consultants to the project. Both individuals are experienced managers of fire investigation units and are certified by the U.S. Fire Administration (USFA) as investigators for USFA's major fires investigation project. They helped conduct field interviews, analyzed and documented the findings, and contributed to project reports. TriData President, Philip Schaenman, Vice President, Barbara Pendergist and Research Associate Charles Jennings reviewed the Final Report and offered important insight and perspectives. Throughout the whole project Ms. Darlene Harris patiently worked and reworked report drafts and handled many of the administrative details of the project. Ms. Charlene Cullen also assisted with the project. The contribution of all these individuals is sincerely appreciated.

Especially we want to recognize and thank the following people from each of the study and implementation sites who set aside time to meet with us, offered candid remarks, provided new details on what is working and what is not in fire investigation, and in general, welcomed us to their community.

STUDY SITES

<u>Orlando, Florida</u>

F.E. Reynolds	- Fire Chief
Anthony Coschignano	- District Chief and Unit Commander
Laurie Fraser	- Lieutentant (Police)
John Hackett	- Lieutentant (Fire)
Eullas Brinson, Jr.	- Lieutentant (Fire)

Livingston County, Michigan

Dennis DeBurton	- Sheriff
David Morse	- Prosecutor
Edwin Moore	- Sheriff's Detective
Keith E. Froelich	- Fire Chief, Hartland Volunteer Fire Department
Donald C. Wilson	- President, Livingston County Fire Investigators
Keith R. Voight	- Deputy Chief, Hartland Township
Kerry D. Matheny	- Police Evidence Tech. Howell Police Department
John L. Wright	- Fire Chief, Fowlerville Volunteer Fire Department

Wilmington, N.C.

William Farris	- Mayor
Sam Hill	- Fire Chief
E. E. Benton	- Asst. Chief/Fire Marshal
Louis G. Goodrum	- Master Detective
H. D. Brown	- Lt. Fire Inspector/Investigator II

Rochester, N.Y.

Leonard Huether	- Fire Chief
William Kelly	- Captain
Carl Clawson	- Investigator
William Stacy	- Investigator
Steve Digennaro	- Investigator
Jerold Bills	- Investigator
David Marsh	- District Attorney

IMPLEMENTATION SITES

Norfolk, Virginia

Tom Gardner	- Fire Chief
Herbert C. Redfield Sr.	- Captain, Commanding Officer
Philip H. East	- Investigator
John Koenig	- Investigator
Kenneth R. Harlan Jr.	- Investigator
Doug Palladino	- Investigator
Forest L. Parhan	- Investigator
Woodrow Lewis, Jr.	- Assistant Commonwealth Attorney

Gainesville, Florida

Courtland Collier	- Mayor Pro Tem
Richard Williams	- Assistant Fire Chief
Thomas Elwell	- Judge
J. C. Robertson	- NFPA Representative
George Blow III	- Assistant District Attorney
David Cowart	- Lieutenant, Police Department
David Cage	- Investigator
Jackie Herndon	- Investigator

<u>Kitsap County, Washington</u>

Ron Perkerewicz	- Director of Community Development
Debra Kieffer	- Assistant Fire Marshal
David Lynam	- Fire Inspector/Investigator
Jim Shields	- Fire Chief, District #18
Roy E. West	- Fire Chief, District #1
Bill Meigs	- Fire Chief, District #7
Edward Boucher	- Assistant Fire Chief, District #7
Roy Lusk	- Assistant Fire Chief, District #1
C. Danny Clem	- County Prosecuting Attorney
Ray Magerstaedt	- Chief of Detectives
Smed Wagner	- Detective
Bob Nordnes	- Fire Investigator, District #18
Chris Casad	- Assistant Prosecuting Attorney

EXECUTIVE SUMMARY

Ten years ago the U.S. Fire Administration and other federal agencies identified and promoted a special concept in investigating fires and controlling arson: the Arson Strike Force (or Fire Investigation Unit). Some communities called it an arson squad; others added citizen representatives, expanded the mission to include prevention, and called the group an arson task force. The core unit, however, was a team of fire and police investigators who brought their respective expertise to the job of identifying incendiary fires and bringing arsonists to justice. Since the passage of time had seena drop in the federal monies available to encourage local joint police/fire investigation units, USFA wanted to find out how mainstream units were faring, to what extent the bi-agency approach was in use, and how well investigation units were being managed.

USFA contracted with TriData Corporation of Arlington, Viriginia to conduct an in-depth examination of four investigation units from small to medium size jurisdictions; to survey by phone a wide assortment of investigators, Fire Marshals, Fire Chiefs, Sheriffs, prosecutors, etc; and to help three units uncover aspects of their organization and operations needing improvement. Over 100 individuals including mayors, judges, sheriffs, investigators, and others contributed their knowledge and experience. The TriData team included the Project Manager and two seasoned fire investigation unit managers who are also USFA certified investigators.

A multi-stage selection process resulted in the choice of four sites where the team examined the management and organization factors that had led to relative success with the local unit. The sites were: Wilmington, NC; Orlando, FL; Rochester, NY; and Livingston County, MI. The key aspects that contribute to the viability of these units are discussed in the community profiles section of this report.

The project team then applied the lessons from the first phase and worked with Gainesville, FL; Norfolk, VA; and Kitsap County, WA to help them trouble-shoot selected organizational and management problems and to

4

recommend improvements for their operations and arson control measures, utilizing lessons learned from the other communities in the project, where appropriate.

A considerable amount of information was gleaned from the seven sites that were visited. Between six and fourteen individuals were interviewed in each location. Moreover, dozens of individuals who are directly involved in investigation and prosecution talked to us at fire service meetings and over the phone about the issues and trends most affecting the capability of local units to fight the arson battle. From this input, seven major problems were identified:

1. The fully integrated police/fire investigation team is a rarity and the bi-agency approach is losing ground to a single agency unit with cross-trained investigators.

2. Staffing levels often are being cut and workload per investigator sometimes is being increased on the basis of insufficient information.

3. Flat rank structures are causing high turnover in many units and the lack of a career ladder with a balanced number of slots for junior and senior investigators is resulting in many units having all new or all experienced personnel.

4. Standard investigator training requirements are not widely adopted, training courses are not as available to local units as they should be, and unit managers are receiving little if any training in management.

5. Unit managers are not doing enough management. Many investigators are not obtaining feedback on the quality of their investigations and are not receiving annual evaluations. Some are citing management favoritism of one investigator over the others as a problem. Data management is suffering from the failure to use a system that would

enable the unit to track their work, discern arson trends, diagnose procedural problems, and document their success.

6. The drug wars are impacting fire investigation in urban and rural areas alike. Drug lords are setting fire to the premises of their rivals, the drug refining process is highly explosive and flammable and is causing fires, and investigators are facing dangerous, armed felons when they respond to some of these scenes or try to make an arrest.

7. Smaller communities have special concerns and roadblocks that hamper their ability to investigate fires and pursue arson cases.

The project also uncovered some positive signs in fire investigation. Across the country dedication to duty and a conscientious determination to do the best job possible are the rule, in spite of some overwhelming odds. Also, there have been dramatic advances in the development and incorporation of juvenile firesetter prevention and intervention programs. Finally, where police and fire agencies are teaming up to investigate and prosecute incendiary fires they are meeting with success. In these communities the team approach is making the difference between a marginal effort and an outstanding effort.

A VIEW OF MANAGEMENT IN
FIRE INVESTIGATION UNITS

Issues and Trends
for the 90's

I. BACKGROUND

What is the status of local fire investigation units a decade after the concept of combining police and fire expertise into a single unit was widely promoted? On behalf of the fire service the U.S. Fire Administration wanted an answer to this question and others related to improving fire investigation unit management, so they initiated a year-long research project aimed at bringing to light the key factors affecting the management and organization of these units. There was interest from the field in USFA's finding good examples of how management and organization problems were being solved so that the solutions could be shared throughout the fire investigation community.

There were several reasons why the issue of management was chosen for in-depth review. Over the last ten years numerous projects have been undertaken that covered such themes as arson strike forces, guidelines for prosecutors on handling arson cases, and resources to aid the fire investigator. Now, it was necessary to address the big picture of fire investigation unit management issues covering everything from reporting formats to personnel management and interagency coordination. Many concerns were being voiced from the field about the need to examine how investigation units were being structured and managed. As one unit commander from a southwest city noted, "We need more than fancy tools and special schools; without management guidelines there is no structure for fire investigation."

The fire service wanted to know how mainstream investigation units were faring. The original models for creating special police and fire investigation teams had been derived primarily from a few big cities -- and

7

that was ten years ago. Since the late seventies and early eighties federal monies to support USFA's arson prevention and control programs had fallen sharply. Consequently, USFA's ability to promote arson control strategies and provide technical assistance had dropped correspondingly.

Had local jurisdictions been willing and able to maintain joint investigation units on their own? Were the successful model programs still intact? Had officials in small and medium-size communities adopted arson task force policies and programs? And what about rural areas and volunteer investigators; what adaptations did they have to make? Were there any organizational complications they had to contend with that investigation units from more densely populated areas did not?

To help answer these questions, USFA awarded a contract to TriData Corporation after selecting the company from a competitive bid process. The project aimed to review a sample of fire investigation units from communities that reflected the range of sizes and resources of typical U.S. jurisdictions.

A. Project Goals

The U.S. Fire Administration set several project goals: 1) identify successful methods of organization and management procedures of various arson units; 2) learn what key factors make for a successful administrative operation; and 3) present the key factors in such a manner that other departments could apply the procedures to their own fire/arson investigation unit, new or existing. USFA wanted to conduct a thorough review of four units and then to touch base with a wide sample of other local investigation units to get a reading of current trends and problems confronting fire investigation managers and investigators. After collecting a solid body of information from these contacts, USFA then planned to invite fire investigation units to volunteer as the implementation sites where management audits would be conducted for the purpose of enhancing unit operations.

B. Methodology

The search began for four communities that had several strong fire investigation management features and that were reasonably "typical" insofar as population size, extent of arson problem, and so forth. At first the project team set out to find units that were models of organization and management. Soon it became clear that this was an unrealistic objective: no one unit possesses all the desirable features inherent in good management. Also, the team believed that it would be as instructive to study a site that had experienced a few management problems (to learn how the problems impacted operations, personnel, and quality of work) as to track down model investigation units that might not exist.

TriData began by contacting local fire marshals, fire investigators, State Police, State Fire Marshals, and sheriffs for candidate units to establish a working list of potential sites. These were screened for size and type of community; evidence of some success in dealing with arson; structure of the unit and lead agency; relationship with the prosecutor's office; and special features (e.g., high quality anti-arson programs, low turnover rate, and high clearance rate). It also was important that potential sites exhibit a sincere interest in the project and a willingness to cooperate.

Working from a list of about 25 prospective sites, project personnel then moved to the second level of selection procedures. Information received about the units under consideration was double-checked with other sources and then compared. The list was shortened. Then the unit commanders from the remaining sites were contacted directly and asked a series of more detailed questions. Lastly, the USFA Project Officer and the contractor's Project Manager discussed each of the finalists and chose the following sites:

1. Wilmington, North Carolina
2. Orlando, Florida
3. Livingston County, Michigan
4. Rochester, New York

C. Field Protocols

As noted earlier, it was USFA's purpose to examine management and organization factors that led to relative success in the investigation, follow-up, and prosecution of cases involving intentionally set fires. The nuts and bolts of how to investigate a fire was not studied under this project. As such, the field protocols were designed to address organizational structure, police and fire agency cooperation on investigations, management responsibilities, and prosecutor support.

To capture the needed information in a consistent manner among all the sites it was decided to prepare standard questions organized around four categories:

0   Organizational Features
0   Management Features
0   Anti-Arson Programs
0   Prosecutor's Involvement

The questions became the framework around which the consultants structured their on-site interviews with the Fire Chief, fire marshal, fire investigation unit commander, Sheriff, fire investigators, and the prosecutor's office. Copies of the questions are provided at the end of this section. Since the on-site level of effort generally was limited to one consultant working for two days (project funding precluded the use of two-person teams in most cases), it was especially important to be able to "hit the deck running" upon arriving at the host site. By having the data-gathering methods and instruments already in place one could maximize the time available for interviewing key players and reading through supporting material.

While the field questions triggered the collection of most of the site information collected, they were not the only means by which the consultants evaluated the units. In addition to holding targeted

10

finterviews, project personnel requested copies of a wide range of reports and documents, including:

- 0 Hiring announcements
- 0 Standard operating procedures
- 0 Fire investigation reports
- 0 Job descriptions
- 0 Organization chart
- 0 Monthly and annual reports
- 0 Internal memos
- 0 List of equipment
- 0 Job performance evaluation forms
- 0 Training records

The schedule for site work was standardized. First an orientation meeting was held with the chief elected or appointed official of the community, the Fire Chief or Sheriff, and the unit commander. We explained what USFA wanted to accomplish, how the unit was chosen, and what was desired to be achieved.

After the orientation meeting TriData met with the unit commander, followed by separate interviews with the fire investigators and police officers. Then a meeting was held with the prosecutor's office. This schedule reflected the natural flow of an investigation -- beginning with the officials who politically and monetarily sustained the unit, and ending with the agency that tried the arson cases in court. Each site visit ended with a debriefing attended by the unit commander and/or the Fire Chief or Sheriff.

D. Terminology

Lack of consistent terminology was a constant problem. The word arson is legally defined differently from state to state and the connotation varies among local governments. Is a fire set by a curious six-year-old, arson? Is such a fire counted with other set fires? Should one distinguish between child-set (perhaps under 7 years old) and juvenile-set fires? Can a fire be referred to as arson until it is established as such through confession or trial? If a community calls its fire investigation squad an arson unit, then whenever dispatched, are they bound to bring a criminal search warrant unless the owner agrees to the investigation?

There is a strong need to establish clear definitions for the terms arson, incendiary, suspicious, and set fires. Until there is some level of uniformity in the terminology it is difficult to establish the magnitude of the problem, and to make comparisons across communities.

For the purposes of this study we use incendiary to mean any fire that is deliberately set. We also prefer the words, "fire investigation unit" to "arson unit."

In the next section we provide a close-up of the four study sites and describe certain characteristics of the investigation unit and the community it serves.

II.   FIRE INVESTIGATION UNIT PROFILES

Livingston County, Orlando, Wilmington, and Rochester have some similar and some different factors that contribute to their units' success. They have wide variations in the amount of time the arson unit existed, the caseload, the shifts, the management, and a host of other features. They had similarities too, the most notable of which was a sincere, dedicated cadre of investigators in every community.

Below is a synopsis of each site that allows for a quick comparision of the community and some details about the factors that contribute toward success.

Problem areas are addressed along with those of the implementation sites in Section II, "Common Problems" because they rarely were unique to one site.

ORLANDO, FLORIDA

Population: 160,000

Date Unit Organized: February 1982.

Lead Agency: Fire Department

Personnel:
Unit Commander (District Chief)
2 Lieutenants (Fire)
1 Detective (Police)
1 Secretary

Shifts: 4 ten-hour days 8:00 A.M. - 7:00 P.M. then 3 days off. On call after 7:00 P.M. Minimum three hours paid on all call-back requirements.

Caseload: Total of 150-200 incendiary cases per year; represents about one third of all fire investigations.

KEY FACTORS FOR SUCCESS

Structure and Organization

The fire investigation unit has a direct-line-relationship with the Fire Chief's Office, and as such, a clear and separate identify within the Fire Department (see organization chart). The unit prepares and submits its own budget. This method of organization is one of the strongest possible in terms of allocating resources and signaling that the unit has a specialized role within the Department.

Since the unit does not function as a division of the Fire Marshal's Office -- as is the custom elsewhere -- there is no confusion over whether the investigators must divide their time between investigating fires, inspecting properties, reviewing preplans, or carrying out fire prevention education programs. The unit has the full support of the Fire Chief and enjoys considerable respect from the other divisions. Orlando has prepared good job descriptions and evaluation forms for the unit commander and for the investigators; for copies, contact the Orlando Fire Department, Special Investigative Services.

Equipment

The Orlando unit has the tools and vehicles needed to do a thorough job of investigating fires and bomb threats. It was mentioned by several of the investigators that having reliable and sufficient equipment not only helps them do a good job, but contributes to their overall job satisfaction.

Morale

One of the best indicators of a viable unit is the state of morale. In Orlando, the investigators' morale, job satisfaction, and personal esteem were very good. A spirit of cooperation and mutual respect was evident and obviously contributed to the success of the unit.

Police and Fire Involvement

Orlando's is one of the few truly joint police/fire units identified during the site selection process. The excellent cooperation between police and fire is the result of good planning that occured at the beginning. In 1981, then Fire Captain Anthony Coschignano and Police Investigator Laurie Fraser co-authored a concept paper proposing a Police/Fire Arson Task Force. In this well researched and clearly articulated paper they documented the history of fire investigation in the city, described the present structure and its inherent problems, and recommended that an arson task force be created. The paper elaborated on proposed personnel, documented advantages that would accrue, reviewed a budget, and set some goals. A year later the unit was implemented. A copy of Orlando's proposal to create a fire investigation unit is found in the Appendix.

By 1987 another change was proposed and accepted by the Fire Chief and the Police Chief. They concurred that a single unit budget should be implemented to alleviate multiple administration and operational problems which had developed because the unit functioned under two separate programs: the Property Section of the Police Department and the Special Investigation Services program of the Fire Department. The single budget was seen as a way to eliminate having to split requisitions for equipment and supplies between two sources and to more effectively manage the overtime budget. Full budget authority subsequently was transferred to the fire investigation unit commander.

WILMINGTON, NORTH CAROLINA

Population: 58,000

Date Unit Organized: August 1987

Lead Agency: Wilmington Fire Department

Personnel: Fire Marshal - Unit Commander
                       1 Fire Inspector/Investigator II
                       1 Master Detective
                       1 Police Officer II
                       Secretary (Part-Time)

Shifts: 8 hours (8:00 A.M. - 5:00 P.M.,) Monday - Friday.
                       On call every weekend, compensated with comp time.

Caseload: 16 incendiary cases (investigators are assigned
                       other duties besides fire investigations).

KEY FACTORS FOR SUCCESS

Impetus for Creating Unit

The Wilmington Fire Investigation Team is still in its infancy, having only been in existence for one year at the time of the USFA project site visit. One of the factors that bodes well for the unit is that it was formed as a proactive measure to improve the method and quality of fire investigations, not as a reaction to a crisis. Because the special fire investigation team was formed as a general management solution, it is more likely to survive the ups and downs of arson rates and the shifts in the political scene.

Interagency Cooperation

Inter-agency cooperation was excellent from the start. A Fire Investigation Task Force was set up to study the concept of police/fire investigations. Representatives from the Wilmington Fire Department, Wilmington Police Department, Budget Management Department, and the North Carolina Bureau of Investigation reviewed the Charlotte, North Carolina task force plan and were guided by that as a model. The final plan mapped out a truly interagency effort among police, fire, the State, and the City budget office. The plan also identified the roles for a host of support agencies whose cooperation would be required from time to time. These included:

- 0 FBI
- 0 North Carolina Department of Insurance
- 0 County Mental Health Services
- 0 Social Service Agencies
- 0 City Building Standards Department
- 0 City Juvenile Investigator Team
- 0 County Health Department
- 0 Chamber of Commerce
- 0 State Alcohol Law Enforcement
- 0 County Sheriff's Department
- 0 County Fire Marshal
- 0 District Attorney's Office

Written Procedures

Another highlight of Wilmington's unit is that they have written operating procedures. Their Task Force's recommendations include a discussion of the purpose of the unit, call-out procedures and the responsibilities of key staff. In a later communication the Task Force outlined professional qualifications for a fire/arson investigator and proposed how those special qualifications would be represented in the positions of police investigator and fire inspector.

Division of Responsibility

Wilmington has organized their unit differently from Orlando. Whereas Orlando has a centrally managed and funded unit of fully cross-trained personnel, Wilmington separates the functions of police and fire investigators. Fire investigators handle cause and origin determination at the request of the senior officer at the scene. The fire investigators then request assistance from the police investigator as needed, especially if an arrest is to be made. The police investigators report to the criminal investigation section sergeant. The fire investigators do not have power of arrest nor do they carry weapons. This arrangement appears to work very well for Wilmington largely because they defined all the roles and responsibilities from the start. See the organization chart in the Appendix.

Cooperation

Wilmington has taken an organized approach to improving fire investigation and in the process has demonstrated that even smaller communities can successfully integrate police and fire agencies in the effort to solve arson cases. An excellent spirit of cooperation and appreciation for each other's knowledge and responsibilities exists among the investigators. The dedication to duty and level of morale are noteworthy.

ROCHESTER, NEW YORK

Population: 275,000

Date Unit Organized: In 1980 the Fire Department used LEAA monies to organize the unit.

Lead Agency: Fire Department (under the office of Fire Marshal)

Personnel: Captain - Unit Commander
                           Lieutenant - Second-in-Command
                           2 Police Officers - Arson Task Force (ATF)
                           2 Fire Investigators - Arson Task Force
                           3 Fire Investigators - Fire Related Youth (F.R.Y.)
                             Program
                           4 Cause & Origin Fire Investigators (Full time)
                           4 Back-up Cause and Origin Investigators (as-
                             needed)
                           2 Secretaries

Shifts: For ATF and F.R.Y. units: Rotating shifts of
                           8:00 A.M. to 4.00 P.M. for day work; 3:30 to 11:30
                           P.M. for night work; on call after 11:30 P.M. 'til
                           8:00 A.M. Cause and origin investigators' shifts
                           are from 8:00 A.M. to 6:00 P.M. and 6:00 P.M. to
                           8:00 A.M.

Caseload: Approximately 600 incendiary cases annually.

KEY FACTORS FOR SUCCESS

Organization

Superior structure and organization head the list of success factors for this fire investigation unit. Not only does Rochester incorporate police investigators directly into the unit, but they have organized three sub-units with different, special functions. They also have built in a cadre of as-needed investigators who gain experience as assistants and then become the talent pool from which replacements for the Arson Task Force can be drawn. The Organization Chart is presented in the Appendix.

The unit commander (a captain) and his second-in-command (a lieutenant) oversee the following divisions:

o Primary (cause and origin) Investigations - A group of four fire investigators with four back-up investigators perform the first leg of the investigation to determine cause and origin. During the time they are assigned to thi s unit they gain experience with investigation.

o Arson Task Force - Once a fire has been determined incendiary the Arson Task Force takes over the case if juveniles are not involved. There are two police officers and two fire investigators assigned to the Arson Task Force. Working as a team, these investigators each handle between 200-250 cases per year and average about 60 court appearances annually.

o Fire Related Youth Program (F.R.Y.) - Rochester has taken the Fire Department's role in juvenile firesetter programs to an unusually high level. Three fire investigators are assigned to handle all cases involving juveniles. Any time a juvenile 16 years or younger is involved in a fire, the F.R.Y. unit automatically is assigned to the case. Referrals are received from the suppression force as well. In this division, the investigators each handle between 300-350 cases a year, however they appear in court about fifty times a year; because most juvenile cases are plea-bargained or are cleared through confessions. The F.R.Y. Unit has been recognized as one of the best in the country. A copy of the F.R.Y. Program Data Sheet is included at the end of this section. For more information on their program, contact the Rochester Fire Department.

Written Procedures

Rochester does as good a job as any investigation unit with which we are familiar in putting their standard operating procedures into writing. They regularly review and update their policies and procedures while keeping personnel informed of the changes or new developments through interdepartmental memos. Far from being an exercise in chasing paper, management's proclivity toward writing and circulating procedure changes and new policies ensures that all unit personnel are kept up-to-date. This is particularly important in larger units such as Rochester's.

Each investigator is given a bound procedures manual that thoroughly reviews standard operating procedures. The manual is too lengthy to reproduce in this report; however, a copy can be requested directly by contacting the Rochester investigation unit.

Communication

Communication is reinforced not only by frequent procedures memos, but by daily meetings at the shift change. From 3:30 - 4:00 P.M. the departing and arriving investigators meet for a quick review of the day's activities, the status of pending cases, and administrative issues. Sometimes the district attorney's office attends these sessions to ensure ongoing communication between the investigation process and the prosecutorial process.

Report Formats

Rochester supplies its investigators with good report formats covering all phases and types of investigations as well as monthly reporting. This allows for consistency in the way investigations are documented, facilitates the process of collecting the information, and results in uniform investigation reports for the commander to monitor and for the prosecutor to utilize.

Job Description

The job descriptions in addition to generally describing responsibilities, clearly outline the actions each investigator should take with regard to determining cause and origin, collecting evidence, interviewing, and taking photos.

The job descriptions are sufficiently detailed that they could be used as the basis for performance reviews.

LIVINGSTON COUNTY, MICHIGAN
(Detroit area)

Population:	100,000
Date Implemented:	1968
Lead Agency:	Livingston County Sheriff's Department
Personnel:	Volunteer Fire Chief assigned as Unit Commander 1 Detective, Sheriffs Bureau 5 Support Detectives 23 volunteer firefighters county-wide, recommended by their Fire Chiefs.
Shift:	On call 24 hours/day. Detectives available 8:00 A.M.- 4:00 P.M. per regular shifts, then rotate a 2:00 p.m. - 1O:OO shift every fifth week; after 10:00 p.m. detectives are on call. Volunteers available as needed.
Caseload:	Caseloads vary widely among the detectives and among the volunteer investigators. County-wide they investigate approximately 30 fires per year.

KEY FACTORS FOR SUCCESS

Involvement of Volunteer Fire Departments

Livingston County faced what many counties with volunteer fire departments face -- the need to provide fire investigation services over a large area but on a relatively infrequent basis. In 1968 when the Livingston County Fire Investigation Unit was formed they recruited two firefighters from each volunteer fire department in the County as base personnel to undergo continuous training in fire causation and arson prosecution. These individuals provide immediate response at fire scenes in their localities and advise their Chiefs of the fire cause.

The Unit is under the command of the Sheriff's Department, which is staffed with paid law enforcement officers. A member of the detective bureau is always on call and available to assist in investigations and in procuring legal documents, interviewing witnesses, obtaining evidence, interrogating suspects, and arresting the accused. Since fire suppression personnel and investigators work closely with the detective bureau, one of the detectives has been assigned to serve as liaison between these organizations. The detective has helped maintain a professional working relationship among volunteer firefighters, investigators, and law enforcement personnel. He is sincerely dedicated to the success of the unit and maintains excellent coordination between volunteers and paid personnel. With this system the Sheriff uses the fire service as a vital partner in fire investigation -- rather than relying solely on the detectives to pursue these incidents. See the organization chart in the Appendix for a description of their structure.

Specialization

The unit divides investigation responsibility into cause and origin work and follow-up investigations. This is a practical and successful solution to the challenge of coordinating fire scene investigations between a few full-time paid detectives who are centrally based and a cadre of many volunteer investigators scattered throughout the county. This system also accomodates the training and call out availability differences that exist. For example, a few of the volunteer investigators have participated in a long list of fire and arson investigation training courses over the years, while the newer members of the unit usually bring only minimal, specialized training with them (volunteer investigators are trained after they join the unit). Likewise, within the Sherriff's Office there are different levels of interest and experience in fire investigation. With the detectives, assignments tend to be made to the two or three detectives exhibiting more interest in working with the unit, though all the detectives are expected to assist when necessary.

In the event that a fire incident is deemed incendiary, the Chief or other senior fire officer notifies the Sheriff's Department, which sends a detective to the scene to undertake a full physical fire scene investigation. A fire investigation mobile vn is also dispatched at this time to provide the necessary tools and equipment required to secure evidence.

The volunteer investigators primarily handle the cause and origin analysis; the detectives and the senior volunteer fire investigators do the follow-up investigation, interrogations, evidence collection, and reporting.

Prosecutor Support

The prosecutor's office is considered part of the unit and is one of the factors contributing to the unit's success. This office provides training for members of the unit and offers legal advice on matters which pertain to preservation and collection of evidence, search requirements, crime trends, and current case law pertaining to their authority and use of criminal and arrest warrants. Case preparation and interviewing witnesses is also a function of the prosecutor's office.

When a criminal search warrant is requested the prosecutor's office is notified and the "on-call" prosecutor is briefed on the circumstances of the fire to help determine if a warrant is justifiable. Fatalities are always investigated. The level of cooperation is good and the unit's satisfaction with the prosecutor's office is evident. Only a small percentage of the cases are plea-bargained and then only when a case is particularly weak.

Assistance to Other Jurisdications

A sign of the unit's success is the fact that Livington County unit provides assistance to other Michigan counties interested in combining law enforcement and fire personnel to solve incendiary fires. At least eight other counties in the state have established team concept units with the help of Livingston County fire investigators.

Standard Report Formats

Livingston County's unit has developed some excellent report formats for collecting a wide range of information on different types and different stages of investigation. Space permits the reproduction of only one of these -- the Structure Fire Activity Log (see Appendix) -- but copies of their Structure Fire Worksheet and Vehicle Fire Investigation Format can be requested from the fire investigation unit.

III. FINDINGS

This section presents the major research findings of the study. Fire departments might consider the problems listed here to see if they share any, and to review the positive features and solutions to see whether they would be useful to implement for themselves.

A. COMMON PROBLEMS

1. The joint police/fire investigation concept

The practice of combining police and fire personnel to investigate incendiary fires is on the decline. Budget cuts and the press of other duties has made team investigation units scarce. Even some of the early programs that USFA used to model the task force concept have gone downhill. It was surprising to find that the unit commanders from a few of the better-known units were not willing to recommend their own unit for the project, and thus, were not included in the study. They cited investigator burnout, political changes, and a lack of momentum from Federal agencies on down as part of the reason for the loss of commitment to the bi-agency approach. Many fire investigation units are functioning as fire-only units with police involvement limited to handling arrests.

Certainly there are still some good interjurisdictional units, but the trend has been for the fire department to assume the lion's share of investigation work using cross-trained fire investigators who are also sworn peace officers with powers of arrest. In most of the jurisdictions studied for this project the right to carry a weapon is assigned automatically as a function of arrest powers, though some investigators must do their job bare-handed. Sometimes fire investigators handle the case up to the point of arresting suspects at which time police detectives get involved. In areas served by volunteer fire departments, the sheriff's office, a regional representative from the State Police or State Fire Marshal's office, or a privately-hired investigator investigate fires suspected of being incendiary.

One factor that has affected the move toward single jurisdictional management of fire investigation is the sheer volume of drug-related crimes that are growing exponentially and draining law enforcement agencies of available resources. Strained sometimes to the breaking point with drug and homicide cases, detectives move arson cases further and further down the list of priorities. And since fire investigation tended to be a lower priority among law enforcement agencies anyway, the commitment among police personnel for arson investigation work was not firmly rooted and became transplanted more easily.

The problem is this: incendiary fire cases always have and probably always will require the expertise offered by both those trained in fire behavior, cause, and origin and those trained in crime solving. Wilmington, North Carolina noted in a memorandum proposing a police/fire team that detailed the myriad of qualifications necessary for fire investigation, "Can we expect any one individual to attain all the in-depth knowledge... necessary to do the job? In practicality, the answer must be no... Isn't it more logical to follow the Fire Investigation Team concept?"

If the current trend continues, and it becomes more common for one agency to handle all aspects of investigation, arrest, and case preparation, then it will be necessary to ensure that that one agency has the full spectrum of capabilities necessary to do the job. Even so, there is no real substitute for the years of investigation of fire behavior, cause, and origin experience that the fire agency contributes; nor for the instinct, street knowledge, and network gained from pursuing leads and tracking down suspects that the police agency contributes.

Mary Galvin, a state attorney and successful arson prosecutor in Connecticut, notes, "First and foremost to any successful anti-arson effort is the existence of an arson task force," and "It is critical that police and fire investigators respond immediately to the fire scene." Ms. Galvin notes that when a joint effort is present, prosecutors stand a better chance of successfully trying cases without the "match-in-hand" evidence so frequently demanded by prosecutors before they accept a case.

Unfortunately what we are seeing is that, while joint police and fire agency investigations may still be the ideal, it apparently is just not a feasible approach for a growing number of cities, towns, and counties. Though many fire investigation managers, investigators, prosecutors and others share the belief that joint investigation units are preferable to the singular agency approach, more concrete data is needed to confirm this view. What is needed is a study of arrest, clearance, and conviction rates between communities that investigate fires using police and fire personnel and communities that utilize only fire or law enforcement investigators.

2. <u>Staffing and Workload</u>

Another area of concern mentioned by many study participants was staffing and workload. According to people in the field, management generally is cutting the number of personnel assigned to fire investigation. In volunteer units the reductions are experienced as a by-product of the general decline in the number of new recruits. The lucky units are those that have kept staffing levels constant; none of the units visited reported an increase in personnel.

Has there been a corresponding decrease in intentionally set fires that is driving the reduction in force? TriData reviewed statistics on arson over a three year period for many of the communities participating in this study, and found that fire investigation and arson rates remained the same or increased. These numbers reflected all incendiary fires, not just set fires in structures. Nationally, however, the incidence of arson in structures is going down slowly according to a recent NFIRS report. The report also shows that incendiary and suspicious vehicle fires have risen and that deaths in incendiary and suspicious structure fires went up between 1977 and 1987.

Staffing level decisions typically are being made on the basis of arson incidence data. But this practice is not a good one because consideration must be given to a <u>number</u> of important elements that affect how much work any given unit or investigator accomplishes.

30

First of all, one must ascertain if the investigators in the unit are responsible for origin and cause determination as well as follow-up investigation. In some cases line firefighters do initial origin and cause, calling upon the investigation unit only when accidental causes are ruled out. Also, the more progressive units are doing more than reacting to set fires -- they are spending time planning and conducting prevention and intervention programs. As such, the hours spent counseling curious young firesetters or talking to school groups about the seriousness of lighting fires must be accounted for.

It is important, too, to examine the type of incendiary fire that prevails in the area. Communities with a preponderance of spite and revenge fires usually can count on a higher percentage of confessions and consequently less labor-intensive cases. Where arson-for-profit fires are most numerous, the time per investigation ratio increases as the investigator spends more time untangling paper trails, interviewing witnesses, and so forth. Finally one must ask whether the individuals responsible for fire investigation do that and that only. In other words, are their positions dedicated to investigations, or do they review preplans, respond to fire calls, inspect commercial occupancies, or handle other law enforcement duties?

Many fire investigation unit managers are looking for a set formula that will indicate how many cases per investigator per year will ensure that the investigators are sufficiently challenged while stopping short of overloading the unit. To see whether there was what might be considered a typical or ideal" caseload we conducted a sampling of the caseload of eleven units from different parts of the country. The following chart shows that there is great variability in case loads. Without knowing more about the nature of the caseloads and about the results obtained, one cannot decide to increase or decrease the size of a unit or to add to or subtract from the duties of investigators.

Sample of Investigation Unit Caseload

Fire Department	Number of investigators*	Investigators dedicated to arson only?	# of arson cases/year	Average # of cases per investigator per year
Northwest - rural	7	No	61	9
Rocky Mts. - rural	10	No	19	2
Rocky Mts. - urban	5	Yes	130	13
North Central - urban	2	Yes	162	81
Rocky Mts. - suburb	6	No	12	2
Northeast - suburb/rural	3	No	13	3
Northwest - suburb	5	No	10	2
Southwest - urban	9	Yes	400	44
West Coast - urban	2	Yes	233	117
Southeast - urban	5	Yes	200	40
Southeast - urban	2	No	42	21

*Includes volunteer investigators and units where investigators are tasked with a variety of other duties.

In deciding on an appropriate caseload one must take into consideration the fact that an increasing number of arson units also are responsible for bomb scene investigations, inspections, prevention education, and other tasks. For the most part, bomb-related calls represent only a small portion of the total job, but they need to be factored into the workload. Finally, it is becoming more common for management to assign internal departmental investigations to the fire investigation unit. Again these types of assignments are infrequent, but need to be part of the picture when analyzing workloads and deciding what manpower is needed.

3. Rank Structure

As far as the ranks and titles of fire investigation unit personnel are concerned, there is wide variance here as well. Unit commanders are battalion chiefs, lieutenants, captains, fire marshals, deputy sheriffs, and senior (or chief) fire investigators. The line investigators from fire departments carry titles such as fire or fire cause investigator, arson investigator, deputy fire marshal, inspector, or even deputy fire coordinator. Their counterparts from the police department are known as detectives, deputy sheriffs, or investigators. A few fire departments do not assign new ranks or titles to the firefighter who joins the fire investigation unit. In these cases the firefighters retain that title (firefighter) even though their responsibilities have both changed and increased. From this potpourri of titles and ranks, one thing is clear: there is no standard personnel profile for fire investigation units that is commonly followed. While that alone may or may not be problematic, the lack of career ladders is a detriment to the future of investigation units.

Flat rank structures, in fact, are a big reason why many units experience high turnover. Often the only thing that differentiates an investigator with ten years of experience from an investigator with one is that the former has accumulated more years of cost-of-living increases -- and usually with few if any performance reviews. (We will address job evaluations shortly). To move up one has to move out. So, the community repeatedly loses experienced investigators, bears the cost of training their replacements, and suffers the consequences of arson control being vested in a unit possessing less experience than it otherwise might have.

Lest one conclude that we are implying that an investigator of long-standing is simply one who did not advance, it should quickly be noted that 1) some of these investigators make a conscious decision to stay on because they enjoy investigation work and the schedule, and these advantages outweigh the disadvantages of poor advancement potential, or 2) some senior investigators belong to the rare unit that is structured with different rank and pay levels based on experience and seniority.

In contrast, some units are experiencing the exact opposite problem of those with younger, less experienced investigators. Sometimes the original cadre of fire investigators forms such a good team and over time develops such a comfortable operation that there is no movement out of the unit.

Unfortunately, these units usually find that all the investigators retire within a year or two of each other, taking with them the guts of investigation experience in that community. Unless a manager is forward-thinking enough to plan for this eventuality and/or has the budget to hire and train replacements before their predecessors leave, the unit finds itself in the position of starting from scratch. The presence of a career ladder with a balanced number of slots for each position (even for units with only a few personnel) can help communities avoid the pitfalls of an all senior or an all junior unit.

4. Investigator Training

There are quite a few gaps in training provided to investigators, both before they join the unit and afterward. Most communities try to get their personnel enrolled in at least the 80 hour National Fire Academy course in basic fire investigation. However, many investigators go years before receiving the training; and many departments cannot afford to send even one investigator for training provided at a far away site. In addition there is no universally recognized and adopted curriculum that is tied to certification as a fire investigator. Most investigators have taken a potpourri of courses ranging from non-certificate seminars to full scale, tested training. The National Fire Protection Association and the International Association of Arson Investigators have been addressing the need for standardization and certification. These efforts need to be continued and expanded. Also, as a rule fire investigation unit managers are not spending enough time crafting a package of training programs for their investigators and then ensuring that each investigator takes the training. Until such time as a nationally-recommended curriculum is established, unit managers should inventory the course offerings from the National Fire Academy, the Bureau of Alcohol Tobacco and Firearms, the State Fire Marshal's Office, and others, and select those that meet local needs.

The investigators contacted in this study also cited that structured courses are not as available to local units as they feel is needed. In particular they felt that the National Fire Academy should offer more fire investigation courses as field courses rather than as resident courses. And State Fire Marshals would contribute to more professionalism in fire investigation if they could focus more resources on bringing more training courses to the local units. Cross-training should be promoted and special courses on photographing the scene, collecting and preserving evidence, interviewing witnesses, and courtroom procedures need to be offered routinely.

5. <u>Management</u>

Almost every organization struggles with the dilemma of where to find the time and money to prepare their employees to become managers; how to navigate needed changes through bureaucratic and sometimes hostile channels; and how to "sell" the employees most affected on the changes/improvements. Many of the unit commanders we visited or interviewed by phone stated that management training is needed, and we concur. Some fire investigation unit managers have reached an impasse and are calling for the state and federal government to recognize the dearth of available management courses.

If there is a "last place" on the list of budget priorities at the local level it seems to be earmarked for management training. Local government budgets just cannot and are not setting aside the money to train their managers. "Downsizing" is all the rage and is spreading quickly as the city manager's preferred method of controlling costs. In a host of local government offices nationwide people and information are being managed in a somewhat hit or miss fashion.

Some fire investigation unit managers happen to be naturally good at management and demonstrate common sense in managing their division. They do a respectable job of self-training, read management articles, and communicate well with their people. Others are technically excellent investigators with a long and strong background in fire or police work -- but are not necessarily born managers.

It was found that fire investigation unit managers often are not doing enough managing! A description of the situation commonly encountered in two major categories of management -- people and data -- tells the story more fully.

a.) <u>Managing and Evaluating Staff</u> - For some reason performance reviews, annual or otherwise, seem to be rarely conducted in fire investigation units. This is a major gap in management's duties. Where the units are comprised mostly of veteran investigators who have the confidence that comes with years of experience, the lack of management review generally is not perceived as a problem. Often the senior investigators would just as soon be allowed to function independently from any oversight or quality control checks anyway. However, most investigators want and would benefit from structured feedback about their performance. The younger investigators, especially, look for constructive assessments of whether they are on track with their cases and how they could improve. Many managers are not reviewing cases often enough nor giving adequate feedback to their personnel.

Some managers operate on the premise that no news is good news. An investigator working for this type of manager only receives feedback when he makes a mistake. It is a basic principle of management that employees need to hear both praise and constructive criticism on a regular basis. When a manager is willing to invest that time the employee gets the message that he and his work are valued, that there are performance standards that he must meet or exceed, and that management cares about what he does and how he does it.

Performance reviews are one of the most important functions of a manager, yet personnel evaluations often are postponed or avoided altogether by fire investigation managers which is causing morale problems in some units. Ironically, we found some excellent evaluation forms on paper but they suffer from lack of use! Why? Quite simply, many managers are not comfortable communicating their assessment of others in writing or verbally. Often they do not relish the idea of being evaluated themselves, either. This is especially true if the manager has forged friendships within the unit and socializes with one or more of the investigators during off hours. Also the manager usually participates directly in investigations, and therefore needs to rely on cooperation from each member of the unit. To the manager it can seem risky to upset the emotional balance of the unit by analyzing and grading the investigators' skills. Yet it is necessary to do just that.

Morale problems generally are not caused because a manager conducted a fair and honest evaluation of an employee's performance; but investigators do become demoralized if they are not sure what is expected of them or how they measure up. A manager's failure to assess everyone's strengths and weaknesses and to hold everyone accountable to the same quality and quantity of effort sows the seeds of serious conflict -- each investigator comes to believe he does more work and a better job than do his colleagues. If this undercurrent of dissatisfaction takes root all the benefits derived from effective training and good equipment can be compromised. Investigation unit managers are not paying attention to performance evaluations -- and they need to.

Another problem mentioned during this study was the very human tendency of managers to show preferential treatment to one particular investigator. That person typically had more opportunities for training, was given the nod to attend special seminars that others were not permitted to attend, had a better vehicle for investigations and so the list goes on. Some managers were unaware they were "playing favorites;" others believed it was not flagrant enough to be a problem.

There are two concerns here. One, the fair-haired boy or girl becomes ostracized from the rest of the group, except when they need him to get to the boss. Second, the resentment that builds among the rest is very real and is destructive to working relationships which can impact on the quality of investigation. Managers need to monitor the way they treat their staff and investigators need to make sure they are performing well.

b.) <u>Managing Data and Reports</u> - Many fire departments and fire investigation units still have problems with data collection and data management, In any given unit one can find a good monthly report, detailed numbers on juvenile-set fires (when, where, how, why, etc.), or an accounting of how many investigator hours were spent in court last year. Missing is a system, logically designed, that starts with identifying the information needed and ends with a reporting of those statistics to the local officials who rule on budget allocations affecting the unit. Fire investigation units are cheating themselves when they fail to adequately document information needed to buttress requests for everything from sniffers to raises. Frankly, it is a mystery how units can survive this era of budget-cutting and accountability without more accurate and detailed data and reports. There were exceptions of course but in the main, fire investigation unit managers need to improve the way they document fire investigation work, incendiary fire patterns and trends, case tracking, and expenditures.

Below is a chart that itemizes some of the data that is helpful to collect and analyze. Fire departments might add or subtract categories from this list, but it is a good place to start. Once the data begins to be collected and reported routinely it is possible to discern trends and to apply the information to management decisions.

Suggested Fire Investigation Unit Data

<u>Investigator Time Profiles</u> - Each investigator to documents weekly and tallies monthly, the hours spent in the following tasks:

<u>Investigation-related</u>

- Origin and case determination
- Follow-up investigation
- Processing evidence
- Preparing reports
- Preparing for court
- Court

<u>Other</u>

- Public education/ information
- Reports
- Anti-arson programs
- Equipment maintenance
- Training
- Weapons qualification

<u>Monthly Reports</u> - For the month and year-to-date:

- Number of fires
- Number of incendiary fires
- Number of accidental fires
- Number of undetermined fires
- Status of investigations

<u>Investigation Case Data</u> - For each case document the following:

- Day of week, time of day
- Motive
- Method
- Age of firesetter
- Dollar loss
- Injuries/fatalities

<u>Measures of Effectiveness</u>

- Number and percent of cases open 6 months or longer
- Number and percent of cases accepted for prosecution
- Number and percent of cases with guilty plea
- Number and percent of cases plea-bargained
- Number and percent of cases resulting in a conviction
- Number and percent of cases suspended
- Number and percent of cases cleared

Three years of data often is considered a good base for trend analysis and statistical conclusions, if the results do not vary dramatically year to year without some ready explanation. One must be careful to account for any particular anomalies in a given year that might skew the numbers (e.g., a serial arsonist sets 10 fires in 3 months and terrorizes a neighborhood before being caught. The average number of set fires in the same community is eight per year).

AIMS -- The Arson Information Management System (AIMS) developed under the auspices of the U.S. Fire Administration is intended to help solve the dilemma of developing good fire investigation data reports and projecting a community's arson-prone areas. A second generation of AIMS now is available and USFA is offering regional workshops on how to use the program. The list of communities that have a copy of AIMS is long; but the list of sites actually using AIMS is short. As with any computer program, getting the software is only half the battle won. Users need to know the features and capabilities of the system, how to manipulate the data, how to generate reports, and so forth. Few units have the time or frankly the motivation to self-train, so there is an urgent need to bring AIMS training closer to the potential user. There are two suggestions which might help:

o A specifically-designed course on the whole realm of fire investigation data, applications, and computerization, offered through the National Fire Academy as part of the fire investigation or management curriculum, and

o Regional workshops sponsored by the U.S. Fire Administration that provide hands-on practice and basic instruction on AIMS.

Not so long ago it was considered a special bonus if investigation units designed an information system that identified what data was needed for decision-making, how the data was to be collected, and in what form information would be reported. It is no longer a luxury but a necessity to produce good data and to use it in managing fire investigation units. The need probably will increase since the FBI recently announced that they are cutting back on special arson reporting and relying instead on statistics maintained at the State level. State and local governments will need to consider using a uniform means of tracking and reporting arson that goes beyond what is captured through NFIRS or NFPA's fire reporting system. AIMS may be the solution.

6. The Drug Wars and Fire Investigation

Rising crime rates associated with illegal drugs are taking their toll on fire investigation units in a number of ways. Virtually every unit -- regardless of size or geographic location -- reported that law enforcement agencies were finding less and less time to assist with arson cases because drug-related crimes and homicides were demanding an increasing share of their detective's time. The problem is especially critical where fire investigators are not empowered to make arrests and must rely on detectives to handle interrogations and arrests. Unacceptable delays in police follow-up are occurring in many jursidictions.

The drug scene is impacting fire investigation in other ways too. Crack houses and drug-processing locations are targets of rival drug lords who torch these facilities to gain turf control or to retaliate. They use fire as a effective weapon to build and defend their illegal drug business. Frequently armed, these individuals pose an especially serious threat to the investigator seeking to ask questions of witnesses or to the police officer attempting to arrest a suspect. Moreover, the very process of refining certain drugs is hazardous; it has the potential of being explosive, toxic, and flammable. Finally, drug users who get careless with sources of heat start fires that endanger the health and lives of civilians as well as the fire and police personnel who respond.

So prevalent is the problem of drugs and their impact on fire investigators that we recommend this issue receive special attention from State Fire Marshals, the U.S. Department of Justice, the U.S. Fire Administration, and the Bureau of Alcohol, Tobacco, and Firearms (Treasury Department). Also, local fire investigation units should consider targeting arson control efforts in drug-prone neighborhoods using code enforcement, public education, and undercover operations as tools to accomplish the job.

7. Special Concerns of Fire Investigation Units in Smaller Communities

Arson tends to be thought of as an urban crime, so most attempts at combatting set fires focus on the agencies and structures common to government as found in moderate and large size cities. But arson is a problem in small towns, suburbs, and rural areas, too and some studies show that it is in these jurisdictions that the incidence of arson actually is increasing. Our research from this project bears out both the seriousness of incendiarism in non-metro areas and the fact that investigation efforts tend to be handled differently than they are in more populated areas. In what ways are the arson problem and the resources to combat it different?

First of all fire investigation responsibility is more diffuse. Whereas in metro areas one agency (usually the fire department) generally designates a special unit of investigators to pursue suspicious fires and then coordinates arrest and prosecution duties with a police agency and the district attorney's office, a rural county typically has a plethora of government agencies with concurrent jurisdiction in fire investigation: the volunteer or perhaps combination Fire Department, the Sheriff's Department, the County Fire Marshal, and usually either a State Police and/or State Fire Marshal's Office. Often the closest city provides assistance too by sharing the time and expertise of an investigator, as needed.

Distance also is a factor in how fires are investigated. Because of the territory covered in more rural areas and the fewer resources available, it can take from twelve hours to a few days before a fire ruled as suspicious undergoes a follow up investigation or is checked out by a law enforcement officer. In the meantime the scene may become contaminated, evidence may be destroyed, and the trail of the arsonist may grow cold. Each successive delay makes the prospect of clearing the case through arrest and conviction less likely. Sometimes people decide just not to bother reporting the fire as suspicious because it will set into motion a futile, dragged out process demanding many hours of their time with negligible results. It is far easier to register the cause as "undetermined" or "accidental."

Then there is the problem of the frequency of cases -- there simply are not as many incendiary fires in rural areas and small towns as in bigger cities, though proportionately, the number of set fires as a percentage of total fires may be actually higher. But because the absolute number of incendiary fires is lower, investigators in less populated areas do not get as much practice in handling investigations as their counterparts in big cities, so it is more difficult for them to hone their skills, refine their procedures, and gain experience. Their cases may have a harder time making it to court because of the lack of experience.

Another variable is training. It is not unusual for urban-based investigators to have had some training before joining the arson unit; then formal and informal training is provided routinely during the first year or two they are on the job. A volunteer investigator or an investigator from a small paid department rarely receives origin and cause training or other investigation courses prior to being signed on to investigate fires. Usually their training is done on the job. There is a strong need to provide investigators everywhere (but especially those in smaller communities) with more training, especially in advanced investigation procedures, evidence collection, photographing the scene, interviewing witnesses, and testifying in court.

Finally, the drug scene is finding its way into rural areas and small cities, too, and affecting the availability of law enforcement personnel to help with fire investigation. Large-scale drug suppliers are discovering that air strips in isolated, unpoliced areas offer the perfect landing spot for their planes loaded with illegal drugs. Once the drugs reach their landing destination, the middle-man traffickers push the drugs locally where there is less competition proportionate to demand; no longer do pushers move all the drugs to the nearest big cities. Most local sheriff's departments are being caught off guard and are not prepared to handle the sudden influx of drug problems. Therefore, they find it necessary to assign most of their personnel to the drug problem, and looking for arsonists becomes an even lower priority than before. The signs indicate that this situation will get worse before it gets better.

B. POSITIVE TRENDS IN FIRE INVESTIGATION UNITS

Earlier in this report the specific positive factors contributing to relative success in four units were detailed. Below are presented three overriding positive trends that were noted among the majority of all the departments studied -- those reviewed in-depth as well as those discussed more briefly by phone or at meetings.

1. Dedication to the Job

It is easy to become jaded and cynical in the field of fire investigation. Investigators break their backs to pursue a case, then the culprit gets off with just a slap on the wrist; or investigators get to the scene to find that overhaul has begun despite instructions to the contrary; or the worst -- the slippery arsonist who has eluded investigators for months turns out to be a member of the fire service, sworn to protect the very population he has endangered. Yet across the country the dedication to duty and conscientious determination to do the best job possible are abundantly in evidence. Even the most discouraged investigators are focusing most of their energy on how to improve the system. What is needed is an infusion of support for these professionals from the federal government on down.   This support can take the form of:

o Advanced and specialized training in investigation procedures, interviewing/interrogating, courtroom procedures, and information management,

o Training and technical assistance for the unit managers in staff and data management,

o Research into new ways to meld police agency expertise into fire investigation units, and

o Standardized training requirements, rank structure, and career ladders.

2. Juvenile Firesetter Prevention and Intervention Programs

Perhaps the most dramatic advance in arson control in the last ten years has been the development and acceptance of programs that intervene with firesetting behavior in children.Where once it was the rare department that offered a juvenile counseling program, now these efforts are found in a host of communities.

The role that fire investigators and firefighters play varies considerably from one program to another.In some cases arson investigators and suppression personnel merely refer troubled youths to other agencies where the actual counseling and follow-up takes place. Elsewhere, uniformed fire employees receive special training and work directly with youth who set fires (even in these cases, though, children exhibiting a need for professional, psychiatric help are referred to a mental health agency or private practitioner with the requisite expertise for handling the more serious cases).In Rochester as noted earlier they created a separate section of the investigation unit to specialize in juvenile firesetter intervention programs. There appear to be three reasons why juvenile firesetting counseling programs have been so widely adopted:

1. Juveniles are setting a significant percentage of fires and fire departments and police agencies can no longer dismiss the situation. While cases of the curious pre-schooler experimenting with matches or lighters make up a portion of this juvenile problem -- a growing proportion of juvenile-set fires are attributed to pre-adolescents and adolescents who light fires to "impress" their peers or for kicks. Most departments are reporting increases in these vandalism fires. Some urban departments have discovered that youth gangs require would-be members to set a fire as a rite of passage. Kids anxious to belong willingly comply. When the problem grows to the point where it can no longer be ignored, communities become more serious about instituting programs to attack the problem.

2. Juvenile firesetter counseling programs work. In a recent research project conducted by TriData on the subject of "Proving Fire Safety Works" we found abundant examples of juvenile programs that have been evaluated for impact and proved successful. Because these programs are producing results, communities are willing to maintain them.

3. The U.S. Fire Administration has been an advocate for juvenile counseling programs. They have sponsored the creation of counseling program prototypes and disseminated these along with examples of successful fire department juvenile programs widely throughout the fire service. These efforts have removed the necessity for fire investigation units and fire education specialists to start from scratch. Consequently, more fire departments have incorporated juvenile firesetter counseling programs into their scope of work. Currently USFA and the Department of Justice are sponsoring research into the key factors for success among numerous well-established juvenile counseling programs.

4. A small number of experts in juvenile fire setter programs have given hundreds of training seminars and talks on how to establish such programs, reaching the grassroots of the fire service.

Cooperation between Fire and Law Enforcement Agencies

The status of cooperation between fire and law enforcement agencies logs in at both extremes of the scale. Among the sites we visited and those we communicated with by phone, the police (or Sheriff) and fire agencies exhibited either excellent or poor rapport and cooperation -- there was very little in between. At its best inter-agency support was firmly entrenched and contributed to an overall successful fire investigation effort. At its worst, lack of cooperation was the root cause of an ineffective system, usually characterized by insufficient (or non existent) follow up after fires were ruled incendiary. In short, where the concept of a bi-agency approach has truly been accepted it is working beautifully. Where it has not, incendiary fires are being treated as fires, not as crimes.

IV. MANAGEMENT REVIEWS IN THREE IMPLEMENTATION SITES

The last step initially planned for this project was to locate a fire investigation unit whose managers and local officials were willing to take a close look at their operations and make some organization and management improvements based on recommendations from TriData. Toward the end of the project year, USFA elected to work with three implementation sites rather than just one. A press release was sent to several publishers of commonly-read fire service publications and to fire investigation organizations announcing USFA's desire to test management and organization criteria in three volunteering units.

Response from the field exceeded original expectations as many grass-roots units claimed that a management review was exactly what was needed to improve interagency coordination and the overall quality of investigations.

TriData sent a short application form to each site that expressed interest (a copy is included at the end of this section) and reviewed each returned questionnaire. Three sites demonstrated particularly strong interest in examining the full range of management issues (as opposed to workload analyses, etc.) and were willing to consider implementing changes. The units USFA selected and the primary issue the consultants addressed in each site were:

Community	Primary Issue
o Norfolk, Virginia	Internal management improvements
o Gainesville, Florida	Budget and administrative improvements
o Kitsap County, Washington	Organizational and inter-agency cooperation improvements

The communities that volunteered to be considered as implementation sites are to be commended. It was one thing to vie for selection as a "model" or study site at the start of the project where the emphasis was on identifying positive factors. It was another to invite an objective, third

47

party look into both strengths and problem areas and to accept recommendations for change.

Each of the three sites hosted a TriData team that examined the unit's strengths and weaknesses using criteria that was in part based on the findings from the four study sites. Recommendations on changes that should be considered for implementation were discussed with local officials who agreed to have USFA evaluate their progress in six months and ascertain the impact of the changes on the unit's operation. The problems we discovered among the three implementation sites have been incorporated with those from the four study sites and presented in Section III of this report.

At the time this report was completed, each of the three sites already had implemented at least several of the project's recommendations. They are using the suggestions for improvements to strengthen their investigation work, and in turn, to serve their communities. This work with the three sites demonstrated that the recommendations based on a consensus of the investigators, chiefs, prosecutors, and researchers who participated in this study, can indeed help local fire investigation arson units to be improved, even in light of real-world constraints.

V. APPENDIX

ATTACHMENTS TO ORLANDO

o Chart of Organization
o Proposal for Police/Fire Arson Task Force

Organizational Chart
Orlando Arson Task Force

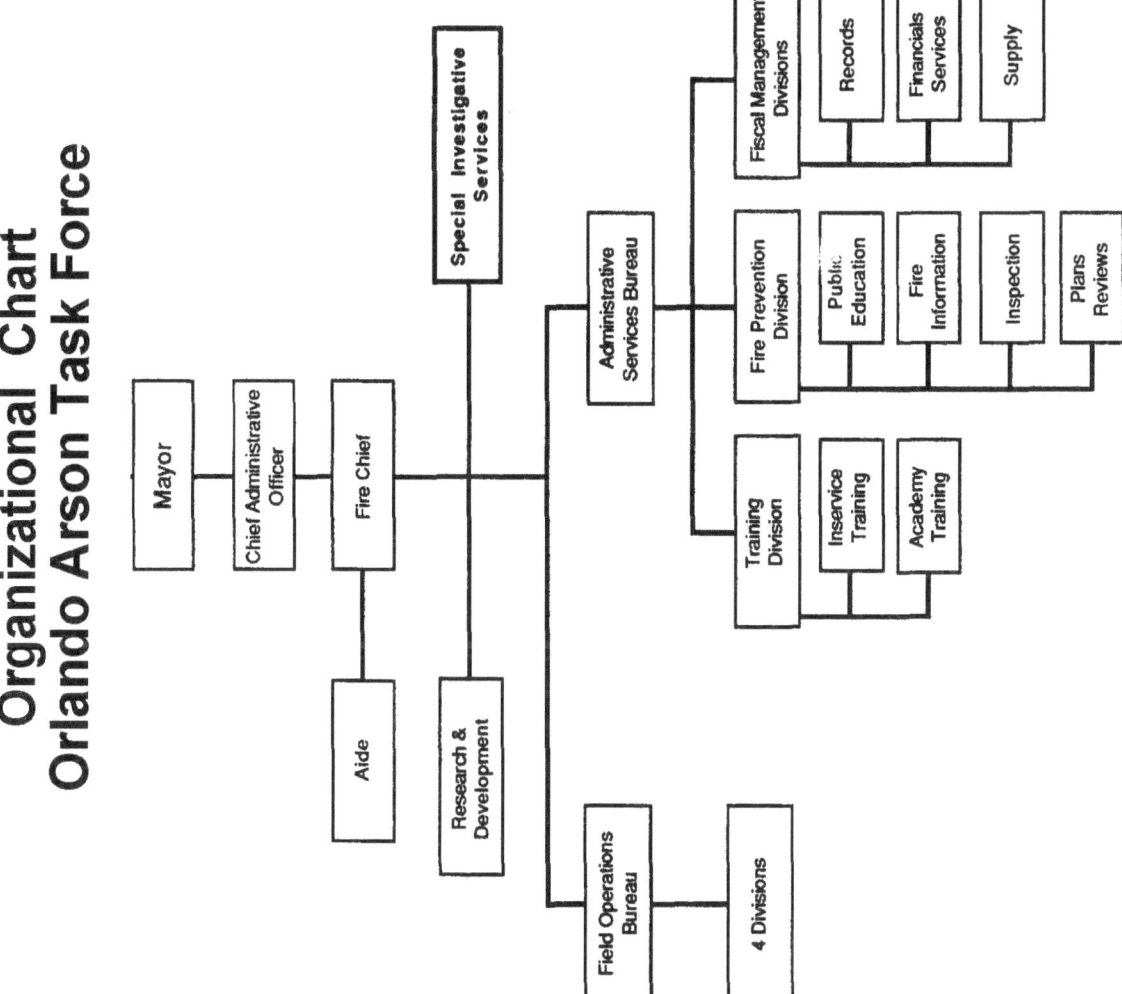

662-10-6-89-17

PROPOSED POLICE/FIRE ARSON TASK FORCE

FOR THE

CITY OF ORLANDO

Prepared by

Captain A.W. Coschignano, OFD
Investigator L.W. Fraser, OPD

INDEX

INTRODUCTION

The crime of arson within the United States has reached epidemic proportions and in 1980 was responsible for a direct loss to property of over 1.7 billion dollars. The indirect loss due to arson; such as, jobs, services, and taxes, amounted to over fifteen billion dollars. The resulting total loss exceeded all of the other property crimes combined. The life loss was over one thousand persons, including fire fighters and other public safety personnel, with ten to 15 thousand people sustaining injuries. Due to the rapidly escalating dollar loss resulting from arson, it has recently been included as a Part 1 Crime in the FBI Reporting Index.

Orlando's arson problem has been held to a level below the national average because of two main factors. First, the economical level of Orlando and surrounding areas has been high and second, arson has been very strongly investigated with a detection, apprehension, and conviction rate which is more than twice that of the national average.

Although these factors have maintained arson in a controlled level within Orlando, the loss factor is increasing at an alarming rate. Orlando's direct loss in 1980 was just under one million dollars. This loss was derived from the insurance claims paid out as a result of arson or suspicious fires. Approximately 27% of the fires within the City of Orlando are of undetermined or suspicious nature and it is estimated that more than 50% of these fires are arson. For 1981 the direct loss for Orlando will exceed one million dollars.

The primary motive for arson throughout the United States has been insurance fraud, followed closely by spite or revenge. In the City of Orlando revenge has been the primary motive, although a significant increase in fraud fires has occurred within recent years with 1981 indicating a record number of fraud or arson for profit cases. This has had the effect of placing an increased burden on the investigative resources within the Police and Fire Department as an arson for profit case requires approximately three to four times the man-hours as most other arson cases. This is due to the extensive examination of the various records required to develop the motive and connect a suspect to the arson.

HISTORY

Prior to 1970, the investigation of fires, not only arson, was almost nonexistent in the City of Orlando. Starting in 1970, it was realized that the investigation of fires, and particularly arson, needed to be conducted in a full time professional manner. A loose team concept was developed in which a fire Department Investigator and a Police Department Investigator teamed up to investigate arson together. The Fire Department Investigator's responsibility was to process the fire scene and if arson was determined then the Police Department Investigator would take over the investigation and work it with the Fire Department Investigator until completion of the case. During the ensuing period of time since the inception of the team concept the total case load has increased from 76 cases in 1971 to a projected total of 668 cases in 1981. During this period of time, two additional Fire Department Investigators have been added, with still only one Police Department Investigator assigned to work arson. Since investigations conducted were of a criminal nature the Fire Department Investigator was sworn as a Special Deputy of the Orange County Sheriff's Department. This was initiated so that he would have a measure of protection during these investigations and was not done with the intention of the Fire Department Investigator conducting the entire investigation. Whenever this team concept was employed in the investigation of arson, the arrest and conviction rate increased significantly.

PRESENT STRUCTURE

Presently there are three Investigators assigned on a permanent full-time basis to the Special Investigative Services Division of the Orlando

Fire and Rescue Department. This Division consists of a Captain who is in charge and reports directly to the Fire Chief, and two Investigators who are Lieutenants and are also on the Chief's Staff. All three Investigators perform the same investigative assignments and are sworn Orange County Special Deputies. The captain also performs administrative duties, which include budget preparation and other reports and evaluations. These Investigators are responsible for the investigation of fires, false fire alarms, false bomb threats, bombing incidents, and internal investigations, as assigned. They are also responsible for inservice training in arson recognition for Operations personnel, counseling of juveniles involved in fire setting and false fire alarms, various public speaking engagements and participation in seminars and training exercises.

The operational structure has remained basically identical to the team concept initially conceived. Although there are three Fire Department Investigators, there is still only one Police Department Investigator assigned to investigate arson. Due to the fact that the Police Investigator is assigned to the Property Section of Criminal Investigation Division, he cannot investigate arson on a full time basis and must carry the additional Property Section case load. This situation has existed for the past five or six years and has caused the Fire Department Investigators to assume a much greater investigative role, which at times included working the entire case from scene to court room. This was not the original intent of the team concept but has evolved out of necessity so that the cases would be worked.

At present, whenever arson is suspected or determined by the Fire Department Investigator who processes the fire scene, the Police Investigator is either requested to respond or is notified of the findings as soon as possible. If the Police Investigator can work the case he and the Fire Department Investigator work as a team. In situations where the Police Investigator cannot break free from his case load or it is a minor case, the Fire Department Investigator will work it to completion. This problem area has caused several cases to be pushed aside or not adequately worked as the Fire Department Investigators have been involved with other assignments or a pressing case load. This also caused a lack of coordination with the Police Department in whose responsibility the investigation of arson as a crime lies. The Fire Department has been forced in many cases to exceed its legal responsibility and conduct criminal investigations of arson. Although the three Fire Department Investigators are sworn Deputies, the legal responsibility of the Fire Department as an agency does not require investigation beyond the fire scene.

PROBLEMS WITH PRESENT STRUCTURE

Not only does the problem exist of not having the Police Investigator available on a full-time basis, but the increasing cases and other work loads of the Fire Department Investigators have reduced their effectiveness in handling arson investigations. Although all three of the Fire Department Investigators work cases and share the case load, the Captain within the Special Investigative Services Division is a supervisor and has the added burden of administrative duties. These duties have increased to the point that a greater percentage of the case load must be handled by the other two Investigators. It must also be realized that arson investigation is only one part of the duties and responsibilities of the Fire Department Investigators. The investigation of accidental fires at times consumes as much or more time than arson investigations. Each year more time is being required of the Fire Department Investigators in the investigation of accidental fires as more and more accidental fires are resulting in civil actions. Also speaking engagements and in-service training are requiring more and more time of the Fire Department Investigators which, consequently, reduces the amount of time that can effectively be spent on an arson investigation.

A minor problem which has developed under the present operational structure is when an arson occurs involving a juvenile or in conjuction with another crime. Whenever a juvenile is involved or suspected of setting a fire the case is sent to the Youth Section of the Police Department who, lacking expertise in fire investigation, turn the case over

56

to the Fire Department for completion. This bypasses the Police Department Arson Investigator who does not receive notification of these cases. The Fire Department Investigator usually completes these investigations and effects the arrest without police assistance. On some occasions the Police Department Arson Investigator does assist the Fire Department Investigator without any case assignment for these cases. This causes a loss in accountability of the Police Arson Investigator's man-hours and case load.

When arson occurs in conjunction with another crime such as auto theft, often neither the Police Department Investigator or Fire Department Investigators are notified. This occurs because many vehicles are stolen and burned with the case being assigned to the Auto Theft Section and worked solely as an auto theft. The vehicles are not properly processed for the crime of arson and as a result only the auto theft is prosecuted which is a lesser degree felony than the arson.

When a fire occurs involving death or injury, the present Police Department MCI Policy states that the responsibility for investigation falls on the Persons Section of the Police Department. This present Policy causes the duplication of investigative effort and a problem of continuity for the Fire Investigators. On any fire resulting in death or injury the Fire Department Investigators request the Police Department Arson Investigator to assist in the fire scene processing. The Police Department Arson Investigator must in turn request a Persons Section Investigator to respond. The case is then turned over to the Persons Investigator after scene processing is complete. This causes a problem in that the Persons Investigator lacks the expertise necessary to fully understand the technical aspects of what occurred on the fire scene and is, therefore, at a disadvantage in follow-up interviews or interrogations. Even if the Fire Department Investigator continues in the investigation with the Persons Investigator, there is a loss of effectiveness by the Persons Investigator not being thoroughly familiar with the fire scene and what has occurred. On several cases the Fire Department Investigators were not included in the follow-up investigation which caused a total lack of continuity from the scene to completion of the case.

When these types of investigative situations occur there is a loss of accountability and the span of control, particularly of the Fire Department Investigators, is increased by having to work with all of these various investigators.

ARSON TASK FORCE

Because of the problems which have already been outlined above and due to the fact that an effective working relationship has been established between Police and Fire Department Investigators, a joint Police/Fire Arson Task Force is proposed for the City of Orlando.

An effective Arson Task Force should contain the following components and meet certain objectives:

1. An Arson Task Force should feature the team concept previously defined. There should be full integration of Police and Fire officers working together, each one bringing their own expertise and unique training to the investigation.

2. An Arson Task Force should be oriented toward a greater range of areas. Arson Investigators should not stop or be impeded by political boundaries. Arsonists are unconcerned with political subdivisions. Investigators also should be unconcerned and concentrate on the apprehension and conviction of these individuals. An Arson Task Force must work and coordinate with investigators or task forces within other jurisdictions or counties.

3. Close integration and cooperation must be maintained with the State Attorney's Office. A future objective is to have an Assistant State Attorney assigned to the Arson Task Force.

4. Periodic training must be conducted for all Police and Fire personnel in arson recognition. This is essential because these

personnel are the eyes and ears of any effective arson investigative effort.

5. It is also imperative to direct arson awareness programs toward the public in order to gain increased attention to the arson problem and cooperation in reducing it.

PROPOSED PERSONNEL

It is proposed that the Arson Task Force be comprised of the following personnel:

a. OFD   - Three Investigators as follows:

One Captain   - Supervisor - presently assigned to SIS Division

Two Lieutenants - Presently assigned to SIS Division

One Secretary   - Presently assigned to the SIS Division

b. OPD - Present   - One Investigator presently assigned to Arson Investigation.

Future   - One Investigator to be assigned and cross trained.

C. Photographer   - Presently utilized photographers. These individuals have been primary fire scene photogrpahers for over-two years and have been assigned an OFD radio, pager and a vehicle.

To be effective, the Arson Task Force must be under the supervision of one individual who will be held accountable for the investigations conducted by the Task Force. It is, therefore, proposed that the Police Department Investigator be assigned to the Task Force on detached service under the supervision of the Fire Department Captain who will be responsible for the Investigator's case load and evaluation. The Fire

Department Captain will also maintain a close liaison with the Criminal Investigation Division Commander within the Police Department and keep him informed of all case developments. A survey of the major arson task forces throughout the United States indicates that this structure is the easiest and most effective method of managing a joint task force.

ADVANTAGES OF PROPOSED PERSONNEL

The listed personnel are currently in their proposed capacity and would require no initial outlay of funds or budgeting to implement them within the Arson Task Force.

With the designation of the Fire Department Captain as Task Force Supervisor, one individual can be held accountable for the Task Force case load and evaluation of the assigned Investigators. This would eliminate the question of who is responsible for monitoring the progress of ongoing investigations.

LOCATION

It is proposed that the Arson Task Force be located on the second floor of the Municipal Justice Building within the area currently occupied by the OFD Special Investigative Services Division.

This area consists of the following:

    Office - Captain, Supervisor
    Office - Investigators
    A Secretary/Reception Area
    Interview Room
    Evidence Room

At present the Investigators office area is occupied by the Orlando Fire and Rescue Department Research and Development Section. There are current plans to move this Section from this location and should this Proposal be accepted, the two Fire Department Investigators and Police

Department Investigator would move into this office area which can accommodate up to four investigators comfortably. The Interview Room presently occupied by the two Fire Department Investigators would revert back to its original purpose. The Captains' Office, Secretary/Reception Area, and Evidence Room would remain unchanged.

The only expense anticipated to effect this relocation of Investigators would be the cost of moving two telephones and installing a third one within the Investigators Office, with the possibility of installing a third line to cover the increased personnel. Desks, chairs, files and office equipment are presently available and would only require relocation; see attached diagram for office layout.

ADVANTAGES OF LOCATION

The location of the Police and Fire Department Investigators within the same office area would result in the following benefits and advantages:

1. Communication and coordination between Investigators would be greatly improved as the Investigators would see each other on a daily basis and be able to discuss and update each other on all aspects of fire related investigations.

2. At present, Investigators must travel between the Police and Fire Department offices to confer on cases. This has resulted in a loss of time and cohesiveness and has caused a disruption and inconvenience to the other investigators within the Criminal Investigation Division.

   This singular location would ensure cohesiveness and unity in the fire related investigations while eliminating the disruptiveness presently occurring.

3. At present, the Police Department Investigator must type his own supplemental and related case reports or dictate them on cassette

tapes and have a Criminal Investigation Division secretary type them. Relocation to the Fire Department offices would eliminate the need for the Police Department Investigator to type reports as the Special Investigative Services Division secretary would perform this function as is presently done for the Fire Department Investigators. This would allow the Police Department Investigator to devote more time to investigations, thereby, increasing efficiency.

The Criminal Investigation Division secretaries would also benefit from this relocation in that they would no longer be required to type these case reports. This would, therefore, allow them more time to devote to the work load of the other investigators.

4. At present, both the Police and Fire Department Investigators maintain identical case files on arson investigations. Due to the Police Department's Central Record system a third file of case related information is maintained in Central Records, This has resulted in a duplication of paperwork, filing, and photographs, which is not only time consuming, but costly,

A joint ATF would require only one master case file containing all of the notes, reports, statements, photographs, etc., developed during the investigation. Thereby, saving time and money in the duplication of paperwork.

This master case file would be located within the Arson Task Force offices and would not only save record storage space but improve investigative efficiency by providing all of the case information in one central file.

5. Evidence directly relating to the point of origin and cause of a fire is secured by the Fire Department Investigators and placed into the Evidence Room located adjacent to the Special Investi-

gative Services Division offices. Other evidence; such as, documents, or evidence requiring latent printing processing, is generally placed into the Police Department Evidence Room. This results in evidence from a case being split between two separate locations. This has caused several problems; such as, lost time in examining evidence, increased paperwork for Police Department evidence personnel, increased chain of custody and lack of continuity in evidence storage.

The Special Investigative Services Division Evidence Room is specifically designed to store fire related evidence; such as, sample cans, flammable liquids, flammable liquid containers, and any other evidence related to the point of origin and cause of a fire.

This Evidence Room contains an evidence indexing system and is totally secure, with only the three Fire Department Investigators having keys and access to it.

By utilizing a single evidence room specifically designed for fire related evidence, the previously mentioned problems would be eliminated and the burden of responsiblity would be removed from the Police Department and a more accountable chain of custody would be maintained.

6. A final advantage of a singular location would be that cross training would improve and become more effective. The Police Department Investigator's expertise in fire scene processing and the Fire Department Investigators expertise in police procedures and post blast investigation would increase significantly resulting in overall improved efficiency.

COMMUNICATIONS

A major problem in the formation of an arson task force between police and fire agencies is the lack of common communication between investigators. This problem does not exist between the Police and Fire Department Investigators due to the fact that two of the three Fire Department Investigators have dual band portable radios containing not only all four OFD channels but also OPD Channels 1 through 4. A third dual band portable has been ordered and will be assigned to the third Fire Department Investigator. The Police Department Investigator also has been assigned a dual band portable containing OPD Channels 1 through 5, and OFD Channels 1 through three, with a built-in pager on the OPD Channels.

The problem of communication by portable radio has been eliminated as all of the Investigators can be in instant contact. Another benefit of this dual communication ability is that the Fire Department Investigators can request vehicle and record checks directly, coordinate with other Police Department officers and investigators and request whatever assistance is required. The Police Department Investigator has gained the benefit of directly requesting fire Department assistance and coordinating with the various Fire Department units, as needed.

It is proposed that an OFD pager be permanently issued to the Police Department Investigator to be utilized after normal duty hours. All three Fire Department Investigators are available by pager after the normal duty hours which greatly facilitates communication with them regardless of their location. Whenever a Fire Department Investigator is requested to respond to a fire scene the Orlando Fire and Rescue Department Communication Division has almost instant contact through the paging system and does not have to waste time trying to locate the Investigator.

The issue of a Fire Department pager to the Police Department Investigator has several distinct benefits as follows:

1. The Fire Department Investigator working a fire scene is utilizing one of the OFD channels and it is faster and more efficient to

request additional assistance through this communication system rather than having to utilize and burden another communication system.

2. When requested and paged, the Police Department Investigator would be required to acknowledge the page by radio, thus, making the Fire Department Investigator on the scene aware of his response and radio communication between Investigators can be established immediately.

3. All of the Fire Department Investigators and Photographers utilized by the Fire Department are available by pager and, therefore, the issue of a pager to the Police Department Investigator link him with a common communication system and allow as many investigative personnel as necessary be notified to respond to an incident scene.

ADVANTAGES OF COMMUNICATION

The dual band communication ability currently exists, with the exception of the permanently assigned pager, and would require no outlay of funds or budgeting for the portable radios. Although the proposed pager and charger would require a capital outlay, the improved communication ability and effectiveness would outweigh additional cost.

OPERATIONAL PROCEDURES

1. GENERAL PROCEDURES

The Fire Department Investigators within the Arson Task Force will continue to investigate all fires and assist in explosive related incidents. The Police Department Investigator will retain the primary responsibility for conducting arson and explosive related investigations, including the utilization of his EOD skills, as required. The cross training of the Police Department Investigator would also enable him to assist in or conduct other fire related investigations.

A second advantage of cross training would be in the area of explosive work. The nature of blast scene investigation is very similar to that used at a fire scene. The Task Force concept would facilitate cross training of the Fire Investigators for post blast operations. This would reduce the need of calling for assistance from outside police and federal agencies, which we have a lack of control over. This would allow for a better operation of an arson and bomb unit.

1. FIRE SCENE PROCEDURES

The Fire Department Investigators will be responsible for the initial investigation of the fire scene and determination of point of origin and cause. The present 24-hour on call status and notification procedure of these Investigators will remain unchanged.

It is proposed that the Police Department Investigator be issued a Fire Department pager and placed in an on call status after normal duty hours. During regular duty hours he will utilize OPD radio channels. After assigned duty hours the Police Department Investigator will be notified and respond to the incident, upon the request of the assigned Fire Department Investigator. The notification of the Police Department Investigator will be based upon the following criteria:

a. fires of suspicious or incendiary nature

b. a large property loss has occurred

c. injury or a death has resulted

d. a suspect is known, or in custody

e. incident involving discovery or detonation of an explosive

f. any other incident which may involve City liability

After the initial fire scene investigation is determined to be arson, the investigation will be conducted jointly until final disposition.

3. INVESTIGATION REPSONSIBILITIES

To establish responsibility whenever a fire or arson has involved another crime or results in death or injury, it is recommended that a joint investigative team be established when necessary. This team will be made up of members of the Arson Task Force and appropriate police investigators. This team concept will be used from the onset of the scene investigation until judicial conclusion.

The primary responsibility for any investigation will fall within the division of the department responsible for the highest criminal violation.

This would set a defined line of responsibility to ensure a cohesive and unified investigation.

4. REPORTS AND EVALUATIONS

The Fire Department Investigators have the primary responsibility of writing the fire scene investigation report, although occasionally the Police Department Investigator will be required to write this report. The subsequent supplemental reports, statements, and other paperwork will be the responsibility of the case investigators and may be written individually or jointly. The Police Department Investigator will also have the responsibility of writing supplemental reports to the initial OPD Incident Reports will be sent directly to the Arson Task Force Supervisor for assignment to the Police Department Investigator.

Case Management will be overseen by the Police Department Investigator and all reports and related paperwork will be reviewed by the Arson Task Force Supervisor prior to filing in Central Records or with the State Attorney's Office. The Supervisor will also evaluate and rate the performance of the Investigators within the Task Force and submit a report to the Criminal Investigation Division Commander indicating the Police Department Investigator's performance and case status.

5. DUTY HOURS AND ASSIGNMENTS

The Fire Department Investigators currently work a 40 hour week, consisting of four 10 hour days. The days off are staggered to ensure that an Investigator is always available during normal duty hours. Duty hours within the Fire Department Investigative Unit are from 0800 hours to 1900 hours and have enabled the Fire Department Investigators to accomplish more work and conduct interviews with individuals not normally available prior to 1800 or 1900 hours. These Investigators are available on call after normal duty hours, including days off, holidays and weekends. Each Investigator is assigned on a primary call out basis for a two day period, with the other Investigators remaining on back up or secondary call. Changes in on call status are handled between the Investigators and the SIS Supervisor. There have been no problems encountered with this work schedule and it is proposed that this schedule remain unchanged.

The Police Department Investigator currently works a 40 hour week, consisting of five 8 hour days, with normal duty hours between 0800 hours and 1600 hours. This current work schedule would present no problems to the operation of the Arson Task Force and would allow the Police Department Investigator equal contact with each Fire Department Investigator. Future expansion or changes in operation procedures may require an adjustment in this work schedule. For the present, however, it is proposed that this schedule remain unchanged.

ARSON TASK FORCE BUDGET

The current SID Division Program Budget would be used as the Arson Task Force Budget with a proportional amount of funding allocated by the Police Department for additional equipment and supplies. Although the Police Department Investigator would be on detached service and assigned to the Fire Department, his salary, overtime, and vehicle would continue to be funded by the Police Department. Daily work sheets for salary overtime purposes would be completed and forwarded to the Police Department.

How this proportioning of funds between the Police and Fire Department budgets for equipment and supplies will be established at the next Fiscal Budget period.

The only foreseeable and initial expense anticipated as a result of the relocation of the, Police Department Investigator would be the moving of several telephones and the possible installation of a third line. Funds for this telephone adjustment are available under the present budget structure. The offices, equipment and supplies already exist, thereby, eliminating a major capital outlay.

ARSON TASK FORCE GOALS

The following are goals to be achieved by the formation of the joint Police/Fire Arson Task Force:

1. Development of a more effective and efficient investigative unit

2. Fully identify and recognize the magnitude of the arson problem within the City of Orlando so that it may be more effectively controlled

3. Reduction in the arson fire loss

4. Increase the ratio of arrests and convictions

5. Promote public awareness concerning arson

6. Establishment of improved levels of training in arson recognition for police and fire personnel

7. Establishment of an effective intelligence and data collection system

CONCLUSION

The formation of joint police/fire arson task forces throughout the United States has resulted in significant success toward decreasing total dollar loss and increasing arrests and convictions. Therefore, this Proposed Police/Fire Arson Task Force will not only result in more effective and unified investigations but will provide improved service to the citizens of Orlando in arson awareness and loss reduction.

ATTACHMENTS TO WILMINGTON

o Chart of Organization

Wilmington Fire Investigation Task Force

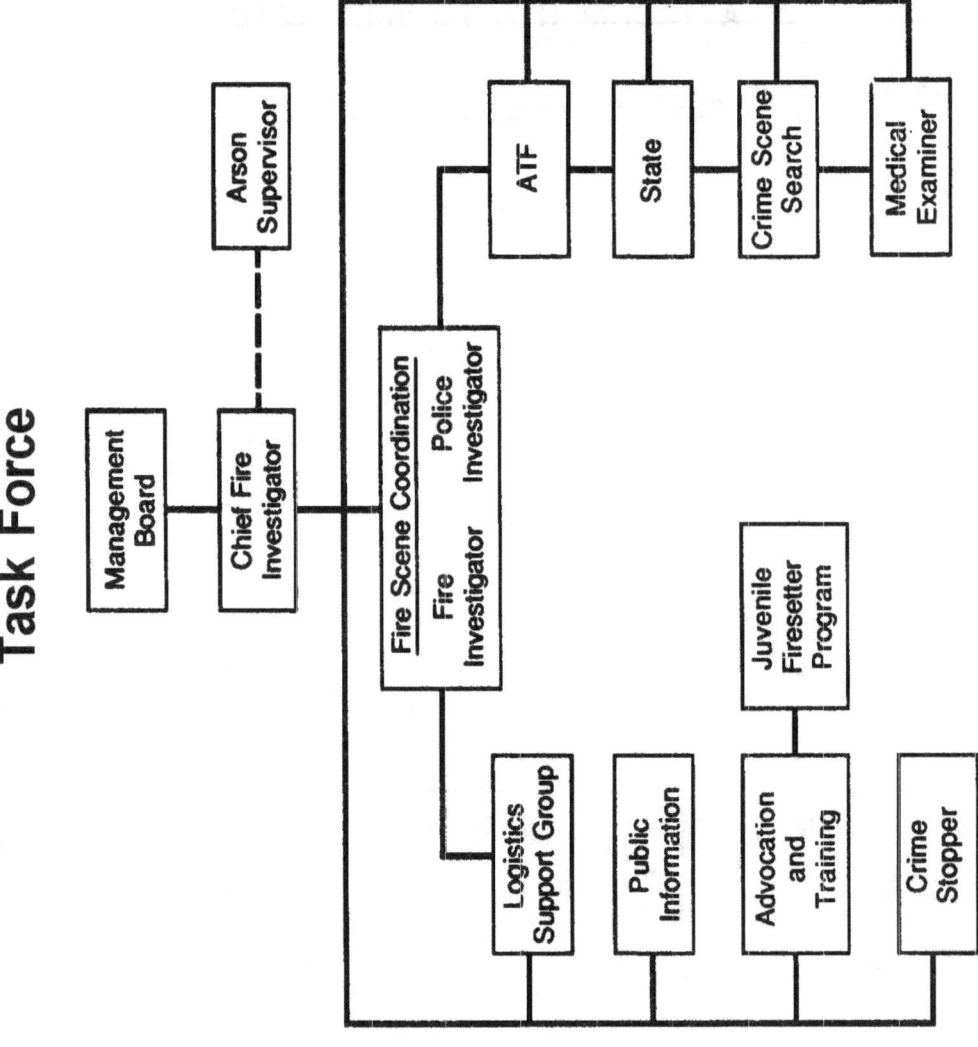

662-10-6-89-16

ATTACHMENT TO ROCHESTER

o Chart of Organization

o FRY Program Data Sheet

Table Of Organization
Rochester Fire Investigation Unit

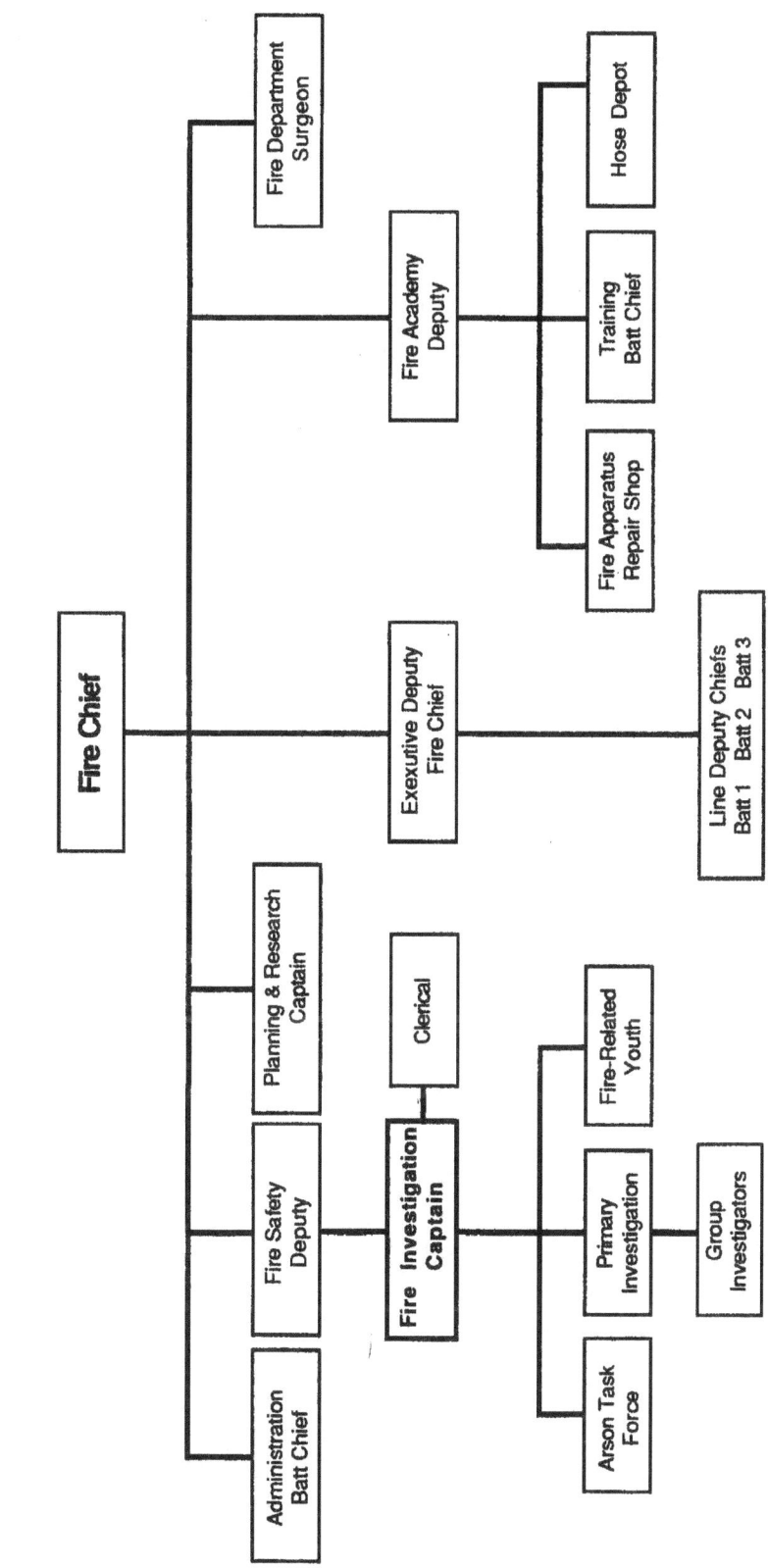

662-10-6-89-14

F.R.Y. PROGRAM DATA SHEET

1. Incident number (from Fire Investigation Report) _____

2. Date of incident __ / __ / __

3. Node number _____

4. Address of incident: _____

5. Was an actual fire set in this incident?   Yes __   No __

6. Was this contact for other than an actual fire?   Yes __   No __

7. False report (telephone in false report)?   Yes __   No __

8. False alarm (pulled alarm box)?   Yes __   No __

INFORMATION REGARDING SUSPECTS

9. Are there identified suspects in this incident?   Yes __   No __

10. Number of children involved.

11. Suspect name and identification:
    (Codes are the first two letters of the first name,
    first two letters of last name.)

Suspect 1: Code _ _ _ _   Last Name _____   First Name _____
Suspect 2: Code _ _ _ _   Last Name _____   First Name _____
Suspect 3: Code _ _ _ _   Last Name _____   First Name _____
Suspect 4: Code _ _ _ _   Last Name _____   First Name _____
Suspect 5: Code _ _ _ _   Last Name _____   First Name _____

12. Are any of the suspects from the same family?   Yes __   No __
    Lists suspects _number_ of children in the family 1 _ _ _ _
                                    in family 2? _ _ _ _
                                    in family 3? _ _ _ _
                                    in family 4? _ _ _ _
                                    in family 5? _ _ _ _

INFORMATION REGARDING FIRE

13.   Referral source: (Circle all that apply)

A = Fire Company
B = Fire Investiqator
        Specify _____
C = Parent or auardians
D = County
E = School
F = Police

G = Dept. of Social Services
H = Mental Health Agency:
        Specify_____
I = Battalion Chief
J = Other:
        Specify_____

14.   Type of fire (Circle all that apply)

A = School
B = Church
C = Vacant lot or street
D = Other unoccupied building, specify _____
E = Car or Truck
F = Mercantile
G = Shed or other building
H = Dumpster or garbage
I = Occupied single family dwelling
J = Occupied multiple family dwelling
K = Other occupied building
L = Neighbor's yard
M = Residential treatment facility
N = Other, specify_____

If occupied, single (I) or multiple (J) family home, circle one of the following:

A = A suspect's bedroom
B = A parent of a suspect's bedroom
C = Sibling of a suspect's bedroom
D = Other bedroom, specify_____
E = Kitchen
F = Bathroom
G = Living room, family room, den, etc.
H = Basement or attic
I = P o r c h
J = Garage

15. Ignition Source (circle all that apply):

A = Matches
B = Lighter
C = Stove
D = Other, specify_____

16. Material or object lit (circle all that apply):

A = Paper, tissue or cardboard
B = Bedding, bed
C = Clothing
D = Toys
E = Candle
F = Leaves, grass, trash
G = Flammable Liquid
H = Firecrackers
I = Furniture
J = Other, Specify_____

17. How were materials obtained: (circle all that apply)

A = Routinely found at home
B = Inadvertently made available
C = Found them
D = Aquired with some effort

18. Was this fire: (circle one)

A = Strictly accidental
B = Result of careless fire play with no intent to
    destroy/damage property or person
C = Result of intent to damage/destroy property or injure person

19. Was there structural damage?   Yes ____ No ____

20. Was this a Code 5?   Yes ____ No ____

21. Was this a multiple alarm?   Yes ____ No ____

22. If someone was injured, fill in number of the following:

A = Juveniles injured   #_____
B = Juveniles burned   #_____
C = Adults injured   #_____
D = Adults burned   #_____
E = Firefighters injured   #_____
F = Firefighters burned   #_____
G = Juvenile fatalities   #_____
H = Adult fatalities   #_____
I = Firefighter fatalities   #_____

CHILD INFORMATION - SUSPECT 1

23. Child identification code (first 2 letters first name, first 2
    letters last name).   _____

75

24. Age of child in years. _____

25. Date of birth __/__/__

26. Sex of child. (F = Female M = Male) _____

27. Race/Ethnicity _____

   A = White
   B = Black
   C = Hispanic
   D = Other   Specify: _____

SCHOOL DATA - SUSPECT 1

28. School grade: _____

   K -12, code grade number or enter:
   SE = Special ed, non-graded class
   RF = Residential facility (e.g. convalescent)
   NS = Not in school

29. If School grade above = SE, circle the following that apply:

   A = Class for learning disabled
   B = For emotionally disturbed
   C = For mentally retared
   D = For physically handicapped

30. Name of school or residential facility: _____

31. Does this child have problems is school (circle one)

   Y = Yes
   N = No

   If yes, circle the following that apply:

   A = Has academic problems (e.g. keeping up grades)
   B = Has been truant from school
   C = Having behavior problems in school

OTHER CHILD CHARCTERISTICS - SUSPECT 1

32. Visible handicap or deformity
    (specify): _____

33. Chronic disability
    (specify): _____

34. Other characteristics: (circle all that apply)

A = Socially isolated
B = Seems (or reported to be) hyperactive.
C = Impulsive
D = Lies or cheats
E = Has stolen
F = Excessive or uncontrollable anger
G = Has been destructive or otherwise violent, destroying others property
H = Is cruel to animals
I = Had prior police contact
J = Child uses alcohol
K = Child abuses drugs

CHILD FIRE INCIDENT INFORMATION - SUSPECT 1

35. N.Y.S. Penal Law Charge: (circle all that apply)

A = Criminal Mischief
B = False Box inside school - (Falsely Reporting)
C = False Box outside school - (Falsely Reporting)
D = False telephone
E = Arson 1
F = Arson 2
G = Arson 3
H = Arson 4
I = Other, specify: _____

36. Has child ever played with matches or ignition materials prior to this occurrence?   Yes _____   No _____

37. Has child set previous fires?
    If yes, answer the following.   Yes _____   No _____

    A.   Aproximately how many?
    B.   Number of prior incidents
         on file with FRY
    C.   Incident number of most
         most recent prior incident   _____

38. Did the fire get out of control?   Yes _____   No _____
    If yes, was the child afraid?   Yes _____   No _____
    Did he attempt to get help?   Yes _____   No _____

39. Does the child now show remorse?   Yes   No _____

40. What was the child's reaction to the fire?
    Does the fire appear as positive or funny

77

to the child?   Yes_____ No_____
Did the child hide?   Yes_____ No_____
Did the child deny responsibility?   Yes_____ No_____
Did the child watch?   Yes_____ No_____

41. Type of firesetting incident (circle one):

     A = Accidental
     B = Curiosity
     C = Emotional
     D = Juvenile Delinquent

   If 41 = B, C, or D, circle all other motives or reasons that apply:

     A = Curiosity about fire
     B = Create excitement
     C = Revenge against (or punish) sibling
     D = Revenge against (or punish) parent
     E = Call attention to own problems
     F = Coercion by friends
     G = Conceal crime
     H = Commit suicide
     I = Response to irresistable urge
     J = Response to unusual idea or fantasy
     K = Response to family difficulties

42. Who was responsible for this child at the time the fire was
     started?   (Circle one)

     A= No one, unsupervised
     B = Older sibling
     C = Adolescent babysitter
     D = Adult babysitter
     E = Parent/Guardian
     F = Other adult
     G = Other (specify)_____

43. Circle all that apply:

     A = Arrest
     B = Child Protective Service
     C = Psychiatric (What Facility?)_____
     D = Shelter
     E = FACIT (R.P.D.)
     F = CARE (R.P.D.)
     G = Youth Service
     H = Caution and Advise
     I = Other (Specify)_____
     J = Juvenile Diversion

44.   Type of family (Circle One)

       A = Two biological parents
       B = Single parent/mother only
       c = Single parent/father only
       D = Stepfamily (either stepmother or stepfather
       E = Adoptive family
       F = Foster family
       G = Mother and other adult
       H = Father and other adult
       I = Other (Specify) _____

45.   Number of children (under 18 years) in family _____

46.   Address of family (if different from incident address)

       _____

47.   Adults living in household:
       Relationship to child   Employed?   Age
       (USE Codes below)   FT   PT   NO

       _____   _____   _____
       _____   _____   _____
       _____   _____   _____

48.   Relationship:

       1 = biological father   8 = foster mother
       2 = biological mother   9 = boyfriend of mother
       3 = stepfather   10 = girlfriend of father
       4 = stepmother   11 = other male relative
       5 = adopting father   12 = other female relative
       6 = adopting mother   13 = other male, specify
       7 = foster father
            14 = other female, specify _____

       _____

49.   Employed FT = Full Time
                PT = Part Time
                NO = Unemployed

50.   If there are family or parent problems use appropriate numbers from
       the following table:

       1 = Yes, investigators observation
       2 = Yes, parental report
       3 = Yes, child report
       4 = Yes, public records or Police records

79

Parent/Guardian indifferent to incident?   ___ ___ ___ ___

Evidence of neglect?
(adult not responsible for child's
welfare)   ___ ___ ___ ___

Any adult hostile to child?   ___ ___ ___ ___

Child abuse?   ___ ___ ___ ___

Conflict among adults?   ___ ___ ___ ___

Adult alcohol abuse?   ___ ___ ___ ___

Adult drug abuse   ___ ___ ___ ___

Parent/guardian subnormal intelligence? ___ ___ ___ ___

Parent/guardian inappropriately angry
or moody?   ___ ___ ___ ___

Parent/guardian exhibit poor contact
with reality?   ___ ___ ___ ___

51.   Any member of household had prior contact with: (circle all that
       apply)

       A = Mental Health Service
       B = Child Protective Service
       c = Police

52.   Does the family receive Public Assistance?   Y e s -   N o -

53.   Does the family provide acceptable climate
       for child? (e.g. reasonably neat, clean,
       adequate size)   Yes   No -

ATTACHMENTS TO LIVINGSTON COUNTY

o Chart of Organization
o Structure Fire Investigation Report
   Format

Organization Chart
Livingston County Fire Investigators

662-10-6-89-15

Vice President
Sherriff Department Liason

Commander in Chief

Chief Fire Investigator

President-Senior
Fire Investigator

Executive Board of Directors

Fire Investigators

Secretary/Treasurer

LIVINGSTON COUNTY FIRE INVESTIGATORS
STRUCTURE FIRE ACTIVITY LOG

1. L.C.S.D. Complaint #_____

2. Other Police Agency and Complaint #_____

3. Fire Department(s) and Run #_____

4. Date and Time of Fire_____

5. Person and agency requesting Investigation_____

6. Date and Time of Investigation_____

7. Time Started_____ Time Finished_____

8. Number of days at the scene_____
   (NOTE: If scene Investigation is more than one day, make out
        a separate activity log with all of due information)

9. Location of Investigation_____

10. Type of Investigation (Fire, Explosion, etc.)_____

11. Item Involved_____

12. Cause of Fire/Explosion_____

13. Insurance Company, Agency, and Amount_____

    _____

14. Investigators _____

    _____

    _____

15. Assisted by (include agency that person is with)_____

    _____

    _____

16. Other_____

    _____

    _____

FIRE SCENE INVESTIGATION REPORT

Incident Date _____

Occupant (Incident location) _____
Address (Incident location) _____
Phone _____

Owner _____

Address _____

City _____

Phone _____

Equipment involved in ignition _____
    Year _____
    Brand Name _____
    Model _____
    Serial No. _____
    Voltage (If any) _____

Mobile Property - Year _____
    Make _____
    Model _____
    V.I.N. _____
    Lic. No. _____

Building Size (Sq. Ft. at Base) _____

Area of Origin _____

Form of Heat Causing Ignition _____

Type of Material First Ignited _____

Use of Material First Ignited _____

Probable Act or Omission _____

_____

Bldg., Veh., Etc. Ins. Co. _____ Amount _____

Contents Insurance Co. _____ Amount _____

Person Making Report _____

Investigation report to the Livingston County Sheriff Department Howell, Michigan.

SUBJECT: Livingston County Sheriff Department

    Complaint Number: _____

    _____
           (Fire Department)
    Fire Incident Number: _____

    _____
      (Other Police Department)
    Complaint Number: _____

INVESTIGATION REQUESTED BY: _____
    (Official title, first and last name)

    _____
    (Agency that person represents)

    Date & Time of Request_____

DATE AND TIME OF INCIDENT: Date Occurred: _____
          (include day of week)
    Time reported: _____

SUBJECT OF INVESTIGATION: _____
    (Type of Investigation, example: car
    fire, dwelling fire, etc.)

    Address:_____
          (Street number and name)

    City or Village_____
      (If within city or village limits)

    Township: _____

    Township Section Number:_____

    County: _____

    State: _____

WHO DISCOVERED INCIDENT: _____

(Name in Full)

_____

(Address - street number, name, city and state)

_____

(TX number) (Date & time-when discovered)

Circumstances of discovery: _____

_____

_____

WEATHER: Skies: _____ Precipitation _____
    Temperature: _____ Wind Direction: _____ Wind speed: _____
    Humidity: _____ Weather Station used: _____

(Station and Location)

Other: _____

_____

(Include any significant changes in weather if different when
scene is processed from info provided by F.D.)

ASSIGNMENT OF INVESTIGATION: _____

_____

_____

(Give list of roads traveled & direction traveled from your location when
notified to the scene. Include time arrived & distance traveled)

OFFICIALS AT SCENE: (Officials at scene upon your arrival, these are to
include persons who are securing the scene. Give
official title with first and last name & agency
that person represents.)

_____   _____

_____   _____

_____   _____

_____   _____

BRIEFING: (Give a brief synopsis of the information you were provided, by witnesses, and/or the official in charge. Include the person's name. Have the Police officer in charge complete an in depth interview.)

_____

_____

_____

_____

SCOPE OF THE INVESTIGATION: A scene search for the purpose of determining the origin and cause of this fire.

OWNER: _____
(Business name if applicable) (name in full of owner) (D.O.B.)

_____
(Owner address)   (City & State)

_____
(Area code & telephone number)

If more than one owner, continue on as above: _____

_____

_____

TENANT: 1. _____   _____
(Business name if applicable) (Business telephone number)

_____
(Tenant's name in full)   (D.O.B.)

_____
(Home telephone No.)

2. _____   _____
(Business name if applicable)   (Business telephone number)

_____
(Tenant's name in full)   (D.O.B.)

_____
(Home telephone No.)

3. _____
    (Business name if applicable)   (Business telephone number

    _____   _____
    (Tenants name in full)   (D.O.B.)

    _____
    (Home telephone No.)

(If more tenants, continue on with listing)

INSURANCE INFORMATION: (If insurance is not known, list as "Unknown at this time." If more than one insurance company, list that information and reason for additional insurance.)

Ins. Agency Name Address (City & State) TX No.	Name of Ins. Co.	Type Policy #
_____	_____	_____
_____	_____	_____
_____	_____	_____

Structure amount:_____ Contents amount:_____

Appurtenant Structure: _____
Other: _____
    (Remarks, recent changes and by whom, additional living expenses, rental car coverage, full replacement policy, etc.)

INVESTIGATORS: (List person(s) in charge first, Fire investigators second, and then other persons who assisted with the scene and what department they were from.)

    TITLE & NAME   DEPARTMENT

_____   _____

_____   _____

_____   _____

_____   _____

_____   _____

_____   _____

PHOTOGRAPHY _____
       (Title, first and last name)

MEASUREMENTS: _____
       (Title, first & last name of person completing diagrams)

       Assisted by: _____
                  (Title, first and last name)

       Assisted by: _____
                  (Title, first and last name)

WORKSHEET: _____
       (Title, first and last name)

_____
       (Title, first and last name)

AUTHORITY TO ENTER: (If more than one consent or more than one type of authority was obtained, list each in their proper chronological order.)

1. _____ Date obtained:_____
   (Ex: Consent, Administrative, etc.)

   Time obtained: _____ Authority given by: _____
                           (Ex: Owner, Tennant, Judge, etc.

   Name: _____ Authority obtained by: _____
                         (Title, Name)

(If investigation stopped for obtaining criminal warrant, or revocation of consent, list time and why. Then list time of re-start and the type of authority to re-enter.)

Date & Time stopped: _____ Reason: _____

_____

_____

INJURIES/FATALITIES(S)

1. Name _____ DOB: _____ TX _____ Address_____

   Pronounced by:_____

_____ Type of injury: _____

Reason at scene: _____

Date: _____ Time: _____ Location: _____

2. Name _____ DOB: _____ TX _____ Address _____

Pronounced by:_____

_____ Type of injury: _____

Reason at scene: _____

Date: _____ Time: _____ Location: _____

3. Name _____ DOB: _____ TX _____ Address _____

Pronounced by:_____

_____ Type of injury: _____

Reason at scene: _____

Date: _____ Time: _____ Location: _____

AREA: (Describe the general area surrounding the incident scene. Ex: rural, residential, subdivision, etc. Include any other important factors that may be necessary.)

_____

_____

_____

_____

INCIDENT SCENE: (Give a brief description of the structure involved. Include type of construction, number of stories, crawl space, basement, etc., type of roof, type of siding, etc. Include the use of the structure, ex: dwelling barn, commercial, etc. Include whether occupied or vacant. Supply dimensions. Include age of structure if known.

_____

_____

_____

_____

UTILITIES: (Describe all types; gas, electric, etc. and who supplied them. Include
where the utilities entered the structure and if they were involved in
the fire or the fire cause. Also, include the type of heating facilities
and whether or not they were involved.)

_____

_____

_____

_____

_____

_____

FIRE DETECTION/SUPPRESSION SYSTEM(S): (If applicable, describe the type of system,
it's location, whether or not they functioned properly, and if they provided
protection in the area of origin. Also, include if the system is monitored by
someone. Ex: alarm company.)

_____

_____

_____

_____

_____

FIRE SCENE EXAMINATION: _____
(Give the date & time the investigation started.)

(Describe the fire damage to the structure starting at the outside and working
towards the interior and to the point of origin, which is the same way the
investigation is conducted. The description should include heat and smoke damage.
Include all burn patterns and "V" patterns. If there are fire victims, include
their location to the area and point of origin. If scene processing takes more than
one visit, or a return visit is made at a later date, the progress of each day's
investigation should be described, as well the date and time each visit is
completed. If the scene is secured between visits, names of persons, dates and
times they secured the scene should be listed.

Include any information that your search of the surrounding area revealed, such as bottles, footwear impressions, tire impressions, etc. Also, include signs of forced entry and by who if known, or if the building was secured.

_____

_____

_____

_____

_____

_____

_____

_____

_____

FIRE SCENE EXAMINATION: (continued)

_____

_____

_____

_____

_____

_____

_____

_____

_____

_____

_____

_____

SPECIAL NOTATIONS: (If applicable, list any and all unusual circumstances and/or conditions found during the investigation. This would include situations or conditions that could not be eliminated as to contributing to the fire, or the cause of the fire. The purpose of this section is to spell out factors that could not be proved or eliminated as to causing or contributing to the cause of the fire.)

_____

_____

_____

_____

_____

_____

_____

_____

COMPLETION OF SCENE EXAMINATION: (Describe the date and time that scene processing was finished.   If the scene was processed on more than one day or one occasion, this should be noted.)

_____

_____

_____

_____

_____

_____

CONCLUSION: (This is the report of the findings of the investigation. Choose the appropriate cause.)

1.   It is the opinion of the Fire Invest.igators that this fire is of an accidental

cause due to_____

_____

_____

2. It is the opinion of the Fire Investigators that this fire was not accidental, but was set by a person or persons unknown.

3. It is the opinion of the Fire Investigators that the cause of this fire is undetermined.

Respectfully submitted,

_____

(Printed name of person making report goes here, signature goes above the line.)
(Title of person making report goes here, example: Fire Investigator, etc.)

_____

☆ U.S. GOVERNMENT PRINTING OFFICE:1992-625.280/60508